Gottlieb and Whitehead Center Groups of Spheres, Projective and Moore Spaces

T0215888

Marek Golasiński • Juno Mukai

Gottlieb and Whitehead Center Groups of Spheres, Projective and Moore Spaces

Springer

Marek Golasiński
Institute of Mathematics
Casimir the Great University
Bydgoszcz, Poland

Juno Mukai
Shinshu University
Matsumoto, Nagano
Japan

ISBN 978-3-319-38454-2 ISBN 978-3-319-11517-7 (eBook)
DOI 10.1007/978-3-319-11517-7
Springer Cham Heidelberg New York Dordrecht London

© Springer International Publishing Switzerland 2014
Softcover reprint of the hardcover 1st edition 2014
This work is subject to copyright. All rights are reserved by the Publisher, whether the whole or part of the material is concerned, specifically the rights of translation, reprinting, reuse of illustrations, recitation, broadcasting, reproduction on microfilms or in any other physical way, and transmission or information storage and retrieval, electronic adaptation, computer software, or by similar or dissimilar methodology now known or hereafter developed. Exempted from this legal reservation are brief excerpts in connection with reviews or scholarly analysis or material supplied specifically for the purpose of being entered and executed on a computer system, for exclusive use by the purchaser of the work. Duplication of this publication or parts thereof is permitted only under the provisions of the Copyright Law of the Publisher's location, in its current version, and permission for use must always be obtained from Springer. Permissions for use may be obtained through RightsLink at the Copyright Clearance Center. Violations are liable to prosecution under the respective Copyright Law.
The use of general descriptive names, registered names, trademarks, service marks, etc. in this publication does not imply, even in the absence of a specific statement, that such names are exempt from the relevant protective laws and regulations and therefore free for general use.
While the advice and information in this book are believed to be true and accurate at the date of publication, neither the authors nor the editors nor the publisher can accept any legal responsibility for any errors or omissions that may be made. The publisher makes no warranty, express or implied, with respect to the material contained herein.

Printed on acid-free paper

Springer is part of Springer Science+Business Media (www.springer.com)

Gottlieb and Whitehead Center Groups of Spheres, Projective and Moore Spaces

Marek Golasiński and Juno Mukai

Abstract

First, we take up the systematic study of the Gottlieb groups $G_{n+k}(\mathbb{S}^n)$ of spheres for $k \leq 13$ by means of the classical homotopy theory methods. We fully determine the groups $G_{n+k}(\mathbb{S}^n)$ for $k \leq 13$ except for the two-primary components in the cases: $k = 9, n = 53$ and $k = 11, n = 115$. Especially, we show that $[\iota_n, \eta_n^2\sigma_{n+2}] = 0$ if $n = 2^i - 7$ for $i \geq 4$.

By use of Siegel's method and the classical results of homotopy groups of spheres and Lie groups, we determine some Gottlieb groups of projective spaces $\mathbb{F}P^n$ over $\mathbb{F} = \mathbb{R}, \mathbb{C}$ or $\mathbb{H}$, the fields of real and complex numbers, respectively, and the skew $\mathbb{R}$-algebra of quaternions, respectively, or give the lower bounds of their orders. Making use of the properties of Whitehead products, we determine some Whitehead center groups of $\mathbb{F}P^n$. Writing $[[-,-]]$ for the least common multiple, we have:

Example 2.13. If $n \geq 2$ then:

1. $P_1(\mathbb{R}P^n) = \frac{1+(-1)^{n-1}}{2}\pi_1(\mathbb{R}P^n)$;
2. $P_2(\mathbb{C}P^n) = \frac{3+(-1)^n}{2}\pi_2(\mathbb{C}P^n)$;
3. $P_4(\mathbb{H}P^n) = [[12, \frac{24}{(24,n+1)}]]\pi_4(\mathbb{H}P^n)$.

Then, the Gottlieb groups $G_{n+k}(M(A,n))$ of Moore spaces $M(A,n)$ for $n \geq 2$ are studied. The groups $G_{n+k}(M(A,n))$ and $G_{n+k}(M(A \oplus \mathbb{Z}, n))$ for $k = 0, 1, 2, 3, 4, 5$, and n odd are determined for a number of abelian finite groups A.

2010 *Mathematics Subject Classification:* primary: 55P05, 55Q15, 55Q40; secondary: 19L20, 55Q50, 55R10.

In memory of Professor Mark Mahowald,
our teacher and intellectual supervisor

Acknowledgements

The authors would like to express their thanks to Department of Mathematics, Dalhousie University, Halifax (Canada), for its hospitality and support on June 29–July 05, 2008.

Further, the second author would like to thank for hospitalities and supports to Faculties of Mathematics and other sciences of Nicolaus Copernicus University, Toruń (Poland), on August 18–31, 2005, and Korea University, Seoul (Korea), on September 25–October 01, 2009.

Contents

List of Tables

Introduction

The Gottlieb groups $G_k(X)$ of a pointed space X have been defined by Gottlieb in [21] and [22]; first $G_1(X)$ and then $G_k(X)$ for all $k \geq 1$. The higher Gottlieb groups $G_k(X)$ are related in [22] to the existence of sectioning fibrations with fiber X. For instance, if $G_k(X)$ is trivial then there is a cross section for every fibration over the $(k+1)$-sphere $\mathbb{S}^{k+1}$, with fiber X. Throughout the volume we do not distinguish between a map and its homotopy class.

Chapter 1 is based on [20] which grew out of our attempt to develop techniques in calculating $G_{n+k}(\mathbb{S}^n)$ for $k \leq 13$ and any $n \geq 1$. The composition methods developed by Toda [85] are the main tools used in the paper. Our calculations also deeply depend on the results of [30, 37, 64].

Section 1.1 serves as background to the rest of this chapter. Write ι_n for the homotopy class of the identity map of $\mathbb{S}^n$. Then, the homomorphism

$$P' : \pi_k(\mathbb{S}^n) \longrightarrow \pi_{k+n-1}(\mathbb{S}^n)$$

defined by $P'(\alpha) = [\iota_n, \alpha]$ for $\alpha \in \pi_k(\mathbb{S}^n)$ [26] leads to the formula $G_k(\mathbb{S}^n) = \ker P'$, where $[-,-]$ denotes the standard Whitehead product. So, our main task is to consult first [26, 28, 43, 45, 84, 85] about the order of $[\iota_n, \alpha]$ and then to determine some Whitehead products in unsettled cases as well. In light of Serre's result [80, Proposition IV.5], the p-primary component of $G_{2m+k}(\mathbb{S}^{2m})$ vanishes for any odd prime p, if $2m \geq k+1$ (Proposition 1.7).

Let EX be the suspension of a space X and denote by $E : \pi_k(X) \to \pi_{k+1}(EX)$ the suspension map. Write $\eta_2 \in \pi_3(\mathbb{S}^2)$, $\nu_4 \in \pi_7(\mathbb{S}^4)$, and $\sigma_8 \in \pi_{15}(\mathbb{S}^8)$ for the Hopf maps, respectively. We set $\eta_n = E^{n-2}\eta_2 \in \pi_{n+1}(\mathbb{S}^n)$ for $n \geq 2$, $\nu_n = E^{n-4}\nu_4 \in \pi_{n+3}(\mathbb{S}^n)$ for $n \geq 4$, and $\sigma_n = E^{n-8}\sigma_8 \in \pi_{n+7}(\mathbb{S}^n)$ for $n \geq 8$. Write $\eta_n^2 = \eta_n \circ \eta_{n+1}$, $\nu_n^2 = \nu_n \circ \nu_{n+3}$, and $\sigma_n^2 = \sigma_n \circ \sigma_{n+7}$. Section 1.2 is a description of $G_{n+k}(\mathbb{S}^n)$ for $k \leq 7$. To reach that for $G_{n+6}(\mathbb{S}^n)$, we make use of Theorem 1.14 partially extending the result of [38]: $[\iota_n, \nu_n^2] = 0$ *if and only if* $n \equiv 4, 5, 7 \pmod 8$ *or* $n = 2^i - 5$ *for* $i \geq 4$; for the proof of which Sects. 1.3 and 1.4 are devoted.

Section 1.5 devotes to proving Mahowald's result: $[\iota_n, \sigma_n] \neq 0$ for $n \equiv 7 \pmod{16}$ and $n \geq 23$.

Section 1.6 takes up computations of $G_{n+k}(\mathbb{S}^n)$ for $10 \leq k \leq 13$ and partial ones of $G_{n+k}(\mathbb{S}^n)$ for $k = 8, 9$ presented in Propositions 1.45–1.48 for $10 \leq k \leq 11$. In a repeated use of [64], we have found out the triviality of the Whitehead product (Mahowald, private communication):

$$[\iota_n, \eta_n^2\sigma_{n+2}] = 0, \text{ if } n = 2^i - 7 \ (i \geq 4),$$

which corrects thereby [64] for $n = 2^i - 7$.

Let X be a connected and pointed space. For $k \geq 1$, define the kth *Whitehead center group* or *P-group* $P_k(X) \subseteq \pi_k(X)$ of elements $\alpha \in \pi_k(X)$ such that the Whitehead product $[\alpha, \beta] = 0$ for all $\beta \in \pi_l(X)$ and $l \geq 1$. Then, $G_k(X) \subseteq P_k(X)$ for all $k \geq 1$ [22]. Furthermore, if X is a H-space, $G_k(X) = P_k(X) = \pi_k(X)$ for all $k \geq 1$.

Let $\mathbb{R}$ and $\mathbb{C}$ be the fields of real and complex numbers, respectively, and $\mathbb{H}$ the skew $\mathbb{R}$-algebra of quaternions. Denote by $\mathbb{F}P^n$ for the n-projective space over $\mathbb{F}$ for $\mathbb{F} = \mathbb{R}, \mathbb{C}$ or $\mathbb{H}$, respectively. Put $d = \dim_{\mathbb{R}}\mathbb{F}$, write $i_{k,n,\mathbb{F}P} : \mathbb{F}P^k \hookrightarrow \mathbb{F}P^n$ for $k \leq n$ the inclusion map and $\gamma_n = \gamma_{n,\mathbb{F}} : \mathbb{S}^{(n+1)d-1} \to \mathbb{F}P^n$ the quotient map. Set $i_{\mathbb{F}} = i_{1,n,\mathbb{F}P} : \mathbb{S}^d \hookrightarrow \mathbb{F}P^n$.

Write

$$O_{\mathbb{F}}(n) = \begin{cases} O(n) & \text{for} \quad \mathbb{F} = \mathbb{R}; \\ U(n) & \text{for} \quad \mathbb{F} = \mathbb{C}; \\ Sp(n) & \text{for} \quad \mathbb{F} = \mathbb{H}, \end{cases}$$

where $O(n)$, $U(n)$, and $Sp(n)$ denote the Lie groups of the orthogonal, unitary, and symplectic $n \times n$ matrices, respectively.

The purpose of Chap. 2 is to determine some P- and Gottlieb groups of $\mathbb{F}P^n$. Our method for the Gottlieb group is based on [39]. That is, we use Siegel's result [81] and the exact sequence induced by the fibration $O_{\mathbb{F}}(n+1) \xrightarrow{O_{\mathbb{F}}(n)\times O_{\mathbb{F}}(1)} \mathbb{F}P^n$.

In the case of the P-groups, we use Barratt–James–Stein's result [8] about the Whitehead products $[\gamma_n\alpha, i_{\mathbb{F}}]$ for $\alpha \in \pi_k(\mathbb{S}^{d(n+1)-1})$.

We also use the results on the homotopy groups of spheres [49,52,53,85,87], the classical groups [37], [46, Appendix A, Table VII–VIII, Topology, pp. 1745–1747], [47,48,50,51,54–57].

Sections 2.1–2.3 expound the necessary notions. Then, we take up the systematic study of the P-groups of $\mathbb{F}P^n$. Writing $[[-, -]]$ for the least common multiple, we have:

Example 2.13. If $n \geq 2$ then:

1. $P_1(\mathbb{R}P^n) = \frac{1+(-1)^{n-1}}{2}\pi_1(\mathbb{R}P^n)$;
2. $P_2(\mathbb{C}P^n) = \frac{3+(-1)^n}{2}\pi_2(\mathbb{C}P^n)$;
3. $P_4(\mathbb{H}P^n) = [[12, \frac{24}{(24,n+1)}]]\pi_4(\mathbb{H}P^n)$.

Section 2.4 presents in Theorem 2.19 the groups $P_{k+n}(\mathbb{R}P^n)$ for $k \leq 13$. The groups $P_{k+2n+1}(\mathbb{C}P^n)$ for $k \leq 14$ are complete and $P_{k+4n+3}(\mathbb{H}P^n)$ for some k determined in Theorems 2.20 and 2.25, respectively, of Sect. 2.5.

Sections 2.6 and 2.7 are devoted to Gottlieb groups $G_k(\mathbb{F}P^n)$. In particular, the groups $G_{k+n}(\mathbb{R}P^n)$ and $G_{k+2n}(\mathbb{C}P^n)$ are partly described. Further, groups $G_{k+4n}(\mathbb{H}P^n)$ for some k are presented in Sect. 2.7 as well.

Motivated by these considerations, for the Cayley projective plane $\mathbb{K}P^2$, the fibration $p : F_4 \longrightarrow F_4/Spin(9) = \mathbb{K}P^2$ is used in Sect. 2.8, to discuss some groups $G_k(\mathbb{K}P^2)$, and presented in Theorem 2.60.

Some particular cases of our results about the Gottlieb groups of $\mathbb{F}P^n$ overlap with those of [40, 77].

The authors of [5, Corollary 4.4] show that the Gottlieb group $G_n(M(\mathbb{Z} \oplus A, n))$ of the Moore space $M(\mathbb{Z} \oplus A, n)$ of type $(\mathbb{Z} \oplus A, n)$ is infinite cyclic for $n \geq 3$ provided n is odd, A is a finite abelian group, and $\mathbb{Z}$ is the infinite cyclic group . The purpose of Chap. 3 is to investigate Gottlieb groups $G_{n+k}(M(A, n))$.

Section 3.1 expounds the necessary notions and results to take up the systematic study of the groups $P_{n+k}(M^n)$ (Proposition 3.10) and derive $G_{n+k}(M^n)$ (Corollary 3.11) for $k = -1, 0, 1, 2, 3, 4$ of mod 2 Moore spaces $M^n = E^{n-2}\mathbb{R}P^2$ for some $n \geq 3$.

Then, Sect. 3.2 concludes with calculations of some Gottlieb groups $G_{m+k}(M^m \vee M(A, n))$ for $k = -1, 0, 1, 2, 3, 4$ (Theorem 3.17) and $G_{n+k}(M(A \oplus \mathbb{Z}, n))$ (Corollary 3.20) for $k = 1, 2, 3, 4, 5$ and a number of finite abelian groups A.

Chapter 1
Gottlieb Groups of Spheres

This chapter published in [20] takes up the systematic study of the Gottlieb groups $G_{n+k}(\mathbb{S}^n)$ of spheres for $k \leq 13$ by means of the classical homotopy theory methods. We fully determine the groups $G_{n+k}(\mathbb{S}^n)$ for $k \leq 13$ except for the two-primary components in the cases: $k = 9, n = 53; k = 11, n = 115$. Especially, we show that $[\iota_n, \eta_n^2\sigma_{n+2}] = 0$ if $n = 2^i - 7$ for $i \geq 4$.

1.1 Preliminaries on Gottlieb Groups

Throughout this paper, spaces, maps, and homotopies are based. We use the standard terminology and notations from the homotopy theory, mainly from [85]. We do not distinguish between a map and its homotopy class.

Let ι_X be the identity class of a connected and pointed space X. Recall that the kth *Gottlieb group* $G_k(X)$ of X has been defined in [21, 22] and is the subgroup of the kth homotopy group $\pi_k(X)$ consisting of all elements which can be represented by a map $f : \mathbb{S}^k \to X$ such that $f \vee \iota_X : \mathbb{S}^k \vee X \to X$ extends (up to homotopy) to a map $F : \mathbb{S}^k \times X \to X$. Define $P_k(X)$ to be the set of elements of $\pi_k(X)$ whose Whitehead product with all elements of all homotopy groups is zero. It turns out that $P_k(X)$ forms a subgroup of $\pi_k(X)$ and, by [22, Proposition 2.3], $G_k(X) \subseteq P_k(X)$. Recall also from [39] that X is said to be a *G-space* (resp. *W-space*) if $\pi_k(X) = G_k(X)$ (resp. $\pi_k(X) = P_k(X)$) for all k.

Given $\alpha \in \pi_k(\mathbb{S}^n)$ for $k \geq 1$, we deduce that $\alpha \in G_k(\mathbb{S}^n)$ if and only if $[\iota_n, \alpha] = 0$. In other words, consider the map

$$P' : \pi_k(\mathbb{S}^n) \longrightarrow \pi_{k+n-1}(\mathbb{S}^n)$$

© Springer International Publishing Switzerland 2014
M. Golasiński, J. Mukai, *Gottlieb and Whitehead Center Groups of Spheres, Projective and Moore Spaces*, DOI 10.1007/978-3-319-11517-7_1

defined by

$$P'(\alpha) = [\iota_n, \alpha] \text{ for } \alpha \in \pi_k(\mathbb{S}^n). \tag{1.1}$$

Then, this leads to the formula

$$G_k(\mathbb{S}^n) = \ker P'.$$

Write now $\sharp$ for the order of a group or its any element. Then, from the above interpretation of Gottlieb groups of spheres, we obtain.

Lemma 1.1. *If $\pi_k(\mathbb{S}^n)$ is a cyclic group for some $k \geq 1$ with a generator α then $G_k(\mathbb{S}^n) = (\sharp[\iota_n, \alpha])\pi_k(\mathbb{S}^n)$.*

Since $\mathbb{S}^n$ is an H-space for $n = 3, 7$, we have

$$G_k(\mathbb{S}^n) = \pi_k(\mathbb{S}^n) \text{ for } k \geq 1, \text{ if } n = 3, 7.$$

We recall the following result from [28, 92] needed in the sequel.

Lemma 1.2. 1. *If $\xi \in \pi_m(X)$, $\eta \in \pi_n(X)$, $\alpha \in \pi_k(\mathbb{S}^m)$, $\beta \in \pi_l(\mathbb{S}^n)$ and if $[\xi, \eta] = 0$ then $[\xi \circ \alpha, \eta \circ \beta] = 0$.*
2. *Let $\alpha \in \pi_{k+1}(X)$, $\beta \in \pi_{l+1}(X)$, $\gamma \in \pi_m(\mathbb{S}^k)$ and $\delta \in \pi_n(\mathbb{S}^l)$. Then $[\alpha \circ E\gamma, \beta \circ E\delta] = [\alpha, \beta] \circ E(\gamma \wedge \delta)$.*
3. *If $\alpha \in \pi_k(\mathbb{S}^2)$ and $\beta \in \pi_l(\mathbb{S}^2)$ then $[\alpha, \beta] = 0$ unless $k = l = 2$.*
4. *$[\beta, \alpha] = (-1)^{(k+1)(l+1)}[\alpha, \beta]$ for $\alpha \in \pi_{k+1}(X)$ and $\beta \in \pi_{l+1}(X)$. In particular, $2[\alpha, \alpha] = 0$ for $\alpha \in \pi_n(X)$ if n is odd.*
5. *If $\alpha_1, \alpha_2 \in \pi_{p+1}(X)$, $\beta \in \pi_{q+1}(X)$ and $p \geq 1$, then $[\alpha_1 + \alpha_2, \beta] = [\alpha_1, \beta] + [\alpha_2, \beta]$ and $[\beta, \alpha_1 + \alpha_2] = [\beta, \alpha_1] + [\beta, \alpha_2]$.*
6. *$E[\alpha, \beta] = 0$ for $\alpha \in \pi_k(X)$ and $\beta \in \pi_l(X)$.*
7. *$3[\alpha, [\alpha, \alpha]] = 0$ for $\alpha \in \pi_{n+1}(X)$.*

Let $G_k(X; p)$ and $\pi_k(X; p)$ be the p-primary components of $G_k(X)$ and $\pi_k(X)$ for a prime p, respectively. Hereafter, we use the results and notations of [85] freely. For $X = \mathbb{S}^n$, recall the notation from [85]:

$$\pi_k^n = \begin{cases} \pi_n(\mathbb{S}^n), & \text{if } k = n; \\ E^{-1}\pi_{2n}(\mathbb{S}^{n+1}; 2), & \text{if } k = 2n - 1; \\ \pi_k(\mathbb{S}^n; 2), & \text{if } k \neq n, 2n - 1. \end{cases}$$

As it is well known,

$$\sharp[\iota_n, \iota_n] = \begin{cases} 1 & \text{for } n = 1, 3, 7, \\ 2 & \text{for odd } n \text{ and } n \neq 1, 3, 7, \\ \infty & \text{for even } n. \end{cases} \tag{1.2}$$

Thus, we have reproved the result [22] that

$$G_n(\mathbb{S}^n) = \begin{cases} \pi_n(\mathbb{S}^n) \cong \mathbb{Z} & \text{for } n = 1, 3, 7, \\ 2\pi_n(\mathbb{S}^n) \cong 2\mathbb{Z} & \text{for } n \text{ odd and } n \neq 1, 3, 7, \\ 0 & \text{for } n \text{ even,} \end{cases} \tag{1.3}$$

where $\mathbb{Z}$ denotes the additive group of integers. It is easily obtained that $G_k(\mathbb{S}^n) = P_k(\mathbb{S}^n)$ for all k, n [39, Theorem I.9]. In other words, on the level of spheres the class of G-spaces coincides with that of W-spaces.

Let now n be odd. Then, by Lemma 1.2: (4), (5) and (7), $[\iota_n, [\iota_n, \iota_n]] = 0$. Furthermore, by Lemma 1.2: (1), (4) and (5), $[2\iota_n, \iota_n] = 0$ implies that $[\iota_n, 2\alpha] = 0$ for any $\alpha \in \pi_k(\mathbb{S}^n)$, that is, $2\pi_k(\mathbb{S}^n) \subseteq G_k(\mathbb{S}^n)$ and thus $G_k(\mathbb{S}^n; p) = \pi_k(\mathbb{S}^n; p)$ for any odd prime p. In light of [41], we also know

$$\sharp[\iota_{2n}, [\iota_{2n}, \iota_{2n}]] = 3, \text{ if } n \geq 2. \tag{1.4}$$

Whence, Lemma 1.2 and (1.4) yield the results proved in [19, Example 3.2].

Corollary 1.3. 1. $(2+(-1)^n)[\iota_n, \iota_n] \in G_{2n+1}(\mathbb{S}^n)$. *In particular, the infinite direct summand of* $G_{4n-1}(\mathbb{S}^{2n})$ *is* $\{3[\iota_{2n}, \iota_{2n}]\}$ *unless* $n = 1, 2, 4$.
2. *If* $k \geq 3$ *then* $G_k(\mathbb{S}^2) = \pi_k(\mathbb{S}^2)$.
3. *If* n *is odd and* $n \neq 1, 3, 7$ *then* $2\pi_k(\mathbb{S}^n) \subseteq G_k(\mathbb{S}^n)$. *In particular,* $G_k(\mathbb{S}^n; p) = \pi_k(\mathbb{S}^n; p)$ *for any odd prime* p *and* $k \geq 1$.
4. $G_k(\mathbb{S}^n) = \pi_k(\mathbb{S}^n)$ *provided that* $E : \pi_{k+n-1}(\mathbb{S}^n) \to \pi_{k+n}(\mathbb{S}^{n+1})$ *is a monomorphism and* $\pi_k^n \subseteq G_k(\mathbb{S}^n)$ *provided that* $E : \pi_{k+n-1}^n \to \pi_{n+k}^{n+1}$ *is a monomorphism.*

We note that P' defined above, the homomorphisms

$$P : \pi_{k+n+1}(\mathbb{S}^{2n+1}) \longrightarrow \pi_{k+n-1}(\mathbb{S}^n)$$

and

$$P : \pi_{k+n+1}^{2n+1} \longrightarrow \pi_{k+n-1}^n$$

in the EHP-sequence denoted by Δ in [85, Chap. II] are related as follows:

$$P' = P \circ E^{n+1}.$$

Now, write $H : \pi_k(\mathbb{S}^n) \to \pi_k(\mathbb{S}^{2n-1})$ for the generalized Hopf invariant [85, (2.7)]. First, we recall the full EHP-sequence:

$$(\mathcal{PE}_{n+k}^n)^+ : \quad \pi_{n+k+2}(\mathbb{S}^{2n+1}) \xrightarrow{P} \pi_{n+k}(\mathbb{S}^n) \xrightarrow{E} \pi_{n+k+1}(\mathbb{S}^{n+1}) \xrightarrow{H} \pi_{n+k+1}(\mathbb{S}^{2n+1})$$

for $k \leq 2n - 1$.

Hereafter, we use often the EHP-sequence of the following type:

$$(\mathcal{PE}^n_{n+k}): \qquad \pi^{2n+1}_{n+k+2} \xrightarrow{P} \pi^n_{n+k} \xrightarrow{E} \pi^{n+1}_{n+k+1} \xrightarrow{H} \pi^{2n+1}_{n+k+1}.$$

It is well known (see, e.g., [92, (2.4)–(2.5) Theorems]) that

$$H[\iota_n, \iota_n] = 0 \text{ for } n \text{ odd, and } H[\iota_n, \iota_n] = \pm 2\iota_{2n-1} \text{ for } n \text{ even.} \tag{1.5}$$

Remark 1.4. Let p be an odd prime. Suppose that $\sharp\alpha = \sharp E^m\alpha = p^l$ for $l, m \geq 1$ and $\alpha \in \pi_{2n+k-1}(\mathbb{S}^{2n-1})$ for $k \geq 0$. Then, $\sharp[\iota_{2n}, E\alpha] = p^l$.

Proof. If $\alpha \in \pi_{2n+k-1}(\mathbb{S}^{2n-1})$ for $k \geq 0$ then, in view of (1.5), we have $H([\iota_{2n}, E\alpha]) = \pm 2E^{2n}\alpha$. Because $\sharp(\pm 2E^{2n}\alpha) = \sharp E^{2n}\alpha = p^l$, we get $\sharp[\iota_{2n}, E\alpha] = p^l$ and the proof follows. □

Let $SO(n)$ be the group of orthogonal matrices and $J : \pi_k(SO(n)) \to \pi_{n+k}(\mathbb{S}^n)$ be the J-homomorphism, and $\Delta : \pi_k(\mathbb{S}^n) \to \pi_{k-1}(SO(n))$ the connecting map associated with the fibration $SO(n+1) \xrightarrow{SO(n)} \mathbb{S}^n$.

Perhaps the following is well known to the experts, but basing on [89] and following [90], we show:

Proposition 1.5. *If* $\alpha \in \pi_k(\mathbb{S}^n)$ *then*

$$J(\Delta\alpha) = \pm[\iota_n, \alpha].$$

Proof. Let G_n be the function space of all maps $\mathbb{S}^n \to \mathbb{S}^n$ with degree one and $F_n = \{f \in G_n;\ f(0, \ldots, 0, 1) = (0, \ldots, 0, 1)\}$. Then, $F_n \cap SO(n+1) = SO(n)$ and, in view of [89, (2.10)], there is an isomorphism (the *Hurewicz isomorphism*)

$$\zeta : \pi_k(F_n) \xrightarrow{\cong} \pi_{n+k}(\mathbb{S}^n).$$

Write $j_n = j : SO(n) \hookrightarrow F_n$ and $\bar{j}_n = \bar{j} : (SO(n+1), SO(n)) \hookrightarrow (G_n, F_n)$ for the inclusion maps. Further, there are natural isomorphisms

$$\tau : \pi_k(SO(n+1), SO(n)) \xrightarrow{\cong} \pi_k(\mathbb{S}^n) \text{ and } \bar{\tau} : \pi_k(G_n, F_n) \xrightarrow{\cong} \pi_k(\mathbb{S}^n) \text{ [89, (2.5)*].}$$

Because $J = \zeta j_*$ and the diagram

$$\begin{array}{ccccc}
\pi_k(\mathbb{S}^n) & \xrightarrow{\tau^{-1}} & \pi_k(SO(n+1), SO(n)) & \xrightarrow{\partial} & \pi_{k-1}(SO(n)) \\
\| & & \downarrow{\scriptstyle \bar{j}_*} & & \downarrow{\scriptstyle j_*} \\
\pi_k(\mathbb{S}^n) & \xrightarrow{\bar{\tau}^{-1}} & \pi_k(G_n, F_n) & \xrightarrow{\bar{\partial}} & \pi_{k-1}(F_n)
\end{array}$$

is commutative, [89, Theorem (3.2)] leads to $J(\Delta\alpha) = \pm[\iota_n, \alpha]$ for $\alpha \in \pi_k(\mathbb{S}^n)$ and the proof is complete. □

By (1.1) and Proposition 1.5, we obtain

$$\ker\{\Delta : \pi_k(\mathbb{S}^n) \to \pi_{k-1}(SO(n))\} \subseteq G_k(\mathbb{S}^n). \tag{1.6}$$

In virtue of [80, Chap. IV] ([85, (13.1)]), Serre's isomorphism

$$\pi_{i-1}(\mathbb{S}^{2m-1}; p) \oplus \pi_i(\mathbb{S}^{4m-1}; p) \cong \pi_i(\mathbb{S}^{2m}; p) \tag{1.7}$$

is given by the correspondence $(\alpha, \beta) \mapsto E\alpha + [\iota_{2m}, \iota_{2m}] \circ \beta$ for $\alpha \in \pi_{i-1}(\mathbb{S}^{2m-1}; p)$ and $\beta \in \pi_i(\mathbb{S}^{4m-1}; p)$.

Certainly, the relation (1.7) implies that $E : \pi_{i-1}(\mathbb{S}^{2m-1}; p) \to \pi_i(\mathbb{S}^{2m}; p)$ is a monomorphism. Further, the EHP-sequence on the p-primary components shows that $E : \pi_{i-1}(\mathbb{S}^{2m-1}; p) \to \pi_i(\mathbb{S}^{2m}; p)$ an epimorphism if and only if $E : \pi_{i-1}(\mathbb{S}^{2m-1}; p) \to \pi_i(\mathbb{S}^{2m}; p)$ is an isomorphism, or equivalently, $\pi_i(\mathbb{S}^{4m-1}; p) = 0$.

To state the next result, we recall from [6, Corollary (7.4)] the following very useful in the sequel:

Proposition 1.6. *If $\alpha \in \pi_m(X)$, $\beta \in \pi_n(X)$ with $m, n > 1$ and $\gamma \in \pi_q(\mathbb{S}^m)$ then*

$$[\alpha \circ \gamma, \beta] = [\alpha, \beta] \circ E^{n-1}\gamma + \sum_{i=0}^{\infty}(-1)^{(i+1)(n+1)}\sigma_{i+1}(\alpha, \beta) \circ E^{n-1}h_i(\gamma),$$

where $h_i(\gamma)$ is the higher Hopf invariant of γ, $\sigma_0(\alpha, \beta) = [\alpha, \beta]$ and $\sigma_{i+1}(\alpha, \beta) = [\alpha, \sigma_i(\alpha, \beta)]$.

Writing $\mathbb{Z}_n\{\alpha\}$ for the cyclic groups of order n generated by α, we show:

Proposition 1.7. *Let p be an odd prime.*

1. *If $p \geq 5$ then*

$$G_i(\mathbb{S}^{2n}; p) = \ker\{E^{2n-1} : \pi_i(\mathbb{S}^{2n}; p) \to \pi_{2n+i-1}(\mathbb{S}^{4n-1}; p)\}.$$

In particular, if $\pi_i(\mathbb{S}^{2n}; p)$ is cyclic with a generator α and then $G_i(\mathbb{S}^{2n}; p) = \{\sharp(\operatorname{im} E^{2n-1})\alpha\}$.

2. *If $E : \pi_{i-1}(\mathbb{S}^{2n-1}; p) \to \pi_i(\mathbb{S}^{2n}; p)$ is an epimorphism then*

$$G_i(\mathbb{S}^{2n}; p) = \ker\{E^{2n-1} : \pi_i(\mathbb{S}^{2n}; p) \to \pi_{2n+i-1}(\mathbb{S}^{4n-1}; p)\}.$$

In particular, if $\pi_i(\mathbb{S}^{2n}; p)$ is cyclic with a generator α and the above is satisfied then $G_i(\mathbb{S}^{2n}; p) = \{\sharp(\operatorname{im} E^{2n-1})\alpha\}$.

3. *If $k \leq 2n$ then $G_{2n+k}(\mathbb{S}^{2n}; p) = 0$.*

Proof. 1. If $p \geq 5$ then Lemma 1.2(7) and Proposition 1.6 imply $[\iota_{2n}, \alpha] = [\iota_{2n}, \iota_{2n}] \circ E^{2n-1}\alpha$ for $\alpha \in \pi_i(\mathbb{S}^{2n}; p)$ provided $p \geq 5$. But, in view of (1.7), the map

$$\pi_{2n+i-1}(\mathbb{S}^{4n-1}; p) \to \pi_{2n+i-1}(\mathbb{S}^{2n}; p)$$

given by the correspondence $\gamma \mapsto [\iota_{2n}, \iota_{2n}] \circ \gamma$ for $\gamma \in \pi_{2n+i-1}(\mathbb{S}^{4n-1}; p)$ is a monomorphism. Hence, we get

$$\begin{aligned}&\ker\{P' : \pi_i(\mathbb{S}^{2n}; p) \to \pi_{2n+i-1}(\mathbb{S}^{2n}; p)\}\\ &= \ker\{E^{2n-1} : \pi_i(\mathbb{S}^{2n}; p) \to \pi_{2n+i-1}(\mathbb{S}^{4n-1}; p)\}.\end{aligned}$$

In particular, if $\pi_i(\mathbb{S}^{2n}; p) = \{\alpha\}$ then

$$\ker\{E^{2n-1} : \pi_i(\mathbb{S}^{2n}; p) \to \pi_{2n+i-1}(\mathbb{S}^{4n-1}; p)\} = \{\sharp(\operatorname{im} E^{2n-1})\alpha\}.$$

2. If $\alpha \in \pi_i(\mathbb{S}^{2n}; p)$ then $\alpha = E\beta$ for some $\beta \in \pi_{i-1}(\mathbb{S}^{2n-1}; p)$. Thus, Lemma 1.2(2) yields that $[\iota_{2n}, \alpha] = [\iota_{2n}, \iota_{2n}] \circ E^{2n-1}\alpha$. Then, as in (1), we get

$$\begin{aligned}&\ker\{P' : \pi_i(\mathbb{S}^{2n}; p) \to \pi_{2n+i-1}(\mathbb{S}^{2n}; p)\}\\ &= \ker\{E^{2n-1} : \pi_i(\mathbb{S}^{2n}; p) \to \pi_{2n+i-1}(\mathbb{S}^{4n-1}; p)\}\end{aligned}$$

and $G_i(\mathbb{S}^{2n}; p) = \{\sharp(\operatorname{im} E^{2n-1})\alpha\}$ provided $\pi_i(\mathbb{S}^{2n}; p)$ is cyclic with a generator α.

3. Since $\pi_{r+s}(\mathbb{S}^r; p) = 0$ for $r \geq 1$ and $s = 0, 1, 2$, the EHP-sequence $(\mathcal{PE}^{2n-t}_{4n-t})^+$ induces the isomorphism $E : \pi_{4n-t}(\mathbb{S}^{2n-t}; p) \to \pi_{4n-t+1}(\mathbb{S}^{2n-t+1}; p)$ for $t = -1, 0, 1$. Further, by the Freudenthal suspension theorem, the map $E^m : \pi_{4n+2}(\mathbb{S}^{2n+2}; p) \to \pi_{4n+m+2}(\mathbb{S}^{2n+m+2}; p)$ is an isomorphism for $m \geq 1$. Hence,

$$E^m : \pi_{2n+k-1}(\mathbb{S}^{2n-1}) \to \pi_{2n+k+m-1}(\mathbb{S}^{2n+m-1}; p)$$

is an isomorphism for $k \leq 2n$ and $m \geq 1$. Consequently, in view of (2), $G_{2n+k}(\mathbb{S}^{2n}; p) = 0$ for $k \leq 2n$ and the proof is complete.

□

Denote by $i_n(\mathbb{R})\colon SO(n-1) \hookrightarrow SO(n)$ and $p_n(\mathbb{R})\colon SO(n) \to \mathbb{S}^{n-1}$ the inclusion and projection maps, respectively. We use the following exact sequence induced by the fibration $SO(n+1) \xrightarrow{SO(n)} \mathbb{S}^n$:

$$(\mathcal{SO}^n_k) \quad \pi_{k+1}(\mathbb{S}^n) \xrightarrow{\Delta} \pi_k(SO(n)) \xrightarrow{i_*} \pi_k(SO(n+1)) \xrightarrow{p_*} \pi_k(\mathbb{S}^n) \longrightarrow \cdots,$$

where $i = i_{n+1}(\mathbb{R})$, $p = p_{n+1}(\mathbb{R})$, and $\Delta\colon \pi_k(\mathbb{S}^n) \to \pi_{k-1}(SO(n))$ the connecting map.

Denote by $V_{n,k}$ the Stiefel manifold consisting of k-frames in $\mathbb{R}^n$ for $k \leq n-1$. We consider the commutative diagram:

$$\begin{array}{ccc}
\pi_k(V_{n+1,1}) & \xrightarrow{\ i_*\ } & \pi_k(V_{2n,n}) \\
\Big\downarrow {=} & & \Big\downarrow \Delta' \\
\pi_k(\mathbb{S}^n) & \xrightarrow{\ \Delta\ } & \pi_{k-1}(SO(n)),
\end{array}$$

where $i\colon V_{n+1,1} \hookrightarrow V_{2n,n}$ is the inclusion and Δ' is the connecting map associated with the fibration $SO(2n) \xrightarrow{SO(n)} V_{2n,n}$.

By [12, Theorem 2], Δ' is a split monomorphism if $k \leq 2n-2$ and $n \geq 13$. So, we have $\sharp(\Delta\alpha) = \sharp(i_*\alpha)$ for $\alpha \in \pi_k(\mathbb{S}^n)$ if $k \leq 2n-2$ and $n \geq 13$. Hence, by (1.6) and [30, Table 2], we obtain the following.

Proposition 1.8. *Let $n \geq 13$. Then, $G_{n+k}(\mathbb{S}^n) = \pi_{n+k}(\mathbb{S}^n)$ for $k = 1,2,8,9$ if $n \equiv 3 \pmod 4$; $G_{n+3}(\mathbb{S}^n;2) = \pi^n_{n+3}$ if $n \equiv 7 \pmod 8$; $G_{n+6}(\mathbb{S}^n) = \pi_{n+6}(\mathbb{S}^n)$ if $n \equiv 4,5,7 \pmod 8$; $G_{n+7}(\mathbb{S}^n;2) = \pi^n_{n+7}$ if $n \equiv 15 \pmod{16}$; $G_{n+10}(\mathbb{S}^n;2) = \pi^n_{n+10}$ if $n \equiv 2,3 \pmod 4$; $G_{n+11}(\mathbb{S}^n;2) = \pi^n_{n+11}$ if n is odd unless $n \equiv 115 \pmod{128}$.*

The notation $\pi_{n+m}(\mathbb{S}^n) = \{\alpha_n\}$ ($\{\alpha(n)\}$, *resp.*) means that there exist some $k \geq 1$ and an element α_k ($\alpha(k)$, *resp.*) $\in \pi_{k+m}(\mathbb{S}^k)$ satisfying $\alpha_n = E^{n-k}\alpha_k$ ($\alpha(n) = E^{n-k}\alpha(k)$, *resp.*) for $n \geq k$. For the p-primary component with any prime p, the notation is available.

Given an odd prime p, in view of [85, Lemma 13.5], there exists $\alpha_{i,p}(3) \in \pi_{2i(p-1)+2}(\mathbb{S}^3;p)$ with $p\alpha_{i,p}(3) = 0$ and $\alpha_{i+1,p}(3) \in \{\alpha_{i,p}(3), p\iota_{2i(p-1)+2}, E^{2i(p-1)-1}\alpha_{1,p}(3)\}_1$ for $i \geq 1$. Write $\alpha_{i,3}(3) = \alpha_i(3)$ and notice that $\alpha_i(n) = E^{n-3}\alpha_i(3)$ generates $\pi_{n+4i-1}(\mathbb{S}^n;3) \cong \mathbb{Z}_3$ for $1 \leq i \leq 5$ and $n \geq 3$.

Recall that by [85, Proposition 13.6, (13.7)], it holds

$$\sharp(\alpha_1(n)\alpha_1(n+3)) = \begin{cases} 3 & \text{for } n = 3,4, \\ 1 & \text{for } n \geq 5. \end{cases} \tag{1.8}$$

In view of [85, Theorem 13.9], there exists an element $\alpha'_3(5)$ such that $3\alpha'_3(5) = \alpha_3(5)$. Denote $\alpha'_3(n) = E^{n-5}\alpha'_n(5)$. Then, $\pi_{2n+12}(\mathbb{S}^{2n+1};3) = \{\alpha'_3(2n+1)\} \cong \mathbb{Z}_9$ for $n \geq 2$.

Write $\{-,-,-\}_n$ for the Toda bracket, where $n \geq 0$ and $\{-,-,-\} = \{-,-,-\}_0$. Next, there exists [85, Lemma 13.8]

$$\beta_1(5) \in \{\alpha_1(5), \alpha_1(8), \alpha_1(11)\}_1 \subseteq \pi_{15}(\mathbb{S}^5;3) \cong \mathbb{Z}_9$$

with $3\beta_1(5) = -\alpha_1(5)\alpha_2(8)$. If $\beta_1(n) = E^{n-5}\beta_1(5)$ for $n \geq 5$ then $\pi_{n+10}(\mathbb{S}^n;3) = \{\beta_1(n)\} \cong \mathbb{Z}_9$ for $n = 5,6$ and $\cong \mathbb{Z}_3$ for $n \geq 7$. By (1.7), the map

$E : \pi_{k-1}(\mathbb{S}^{2m-1};3) \to \pi_k(\mathbb{S}^{2m};3)$ is a monomorphism. Then, by the use of [85, Lemma 13.8, Theorems 13.9, 13.10], we obtain

$$3\beta_1(5) = -\alpha_1(5)\alpha_2(8),\ \sharp\beta_1(5) = \sharp\beta_1(6) = 9,\ \sharp\beta_1(n) = 3 \text{ for } n \geq 7 \text{ and}$$
$$\sharp\alpha_1(2n+1)\beta_1(2n+4) = 3 \text{ for } n \geq 1. \tag{1.9}$$

We recall that $\pi_{n+11}(\mathbb{S}^n;3) = \{\alpha_3(n)\} \cong \mathbb{Z}_3$ for $n = 3, 4$ and that $\pi_{n+11}(\mathbb{S}^n;3) = \{\alpha'_3(n)\} \cong \mathbb{Z}_9$ for $n \geq 5$ ([85, Theorem 13.9]), where

$$3\alpha'_3(n) = \alpha_3(n) \text{ for } n \geq 5. \tag{1.10}$$

Let $\Omega^2\mathbb{S}^{2m+1} = \Omega(\Omega\mathbb{S}^{2m+1})$ be the double loop space of $\mathbb{S}^{2m+1}$ and $Q_2^{2m-1} = \Omega(\Omega^2\mathbb{S}^{2m+1}, \mathbb{S}^{2m-1})$ the homotopy fiber of the canonical inclusion (the double suspension map) $i\colon \mathbb{S}^{2m-1} \to \Omega^2\mathbb{S}^{2m+1}$. Then, the (mod p) EHP-sequence [87, (2.1.3)] or [85, (13.2)] is stated as follows:

$$\cdots \xrightarrow{E^2} \pi_{i+3}(\mathbb{S}^{2m+1}) \xrightarrow{H} \pi_i(Q_2^{2m-1}) \xrightarrow{P} \pi_i(\mathbb{S}^{2m-1}) \xrightarrow{E^2} \pi_{i+2}(\mathbb{S}^{2m+1}) \xrightarrow{H} \cdots. \tag{1.11}$$

By making use of [85, Corollary 13.2], we obtain the generators of the following groups which are all isomorphic to $\mathbb{Z}_3$:

$$\pi_{6m-3}(Q_2^{2m-1};3) = \{i(2m-1)\},$$
where $i_{2m-1}\colon \mathbb{S}^{6m-3} \hookrightarrow Q_2^{2m-1}$ is the inclusion;
$$\pi_{6m}(Q_2^{2m-1};3) = \{a_1(2m-1)\}\ (a_1(2m-1) = i(2m-1)\alpha_1(6m-3));$$
$$\pi_{6m+4}(Q_2^{2m-1};3) = \{a_2(2m-1)\}\ (a_2(2m-1) = i(2m-1)\alpha_2(6m-3));$$
$$\pi_{6m+7}(Q_2^{2m-1};3) = \{b_1(2m-1)\}\ (b_1(2m-1) = i(2m-1)\beta_1(6m-3)). \tag{1.12}$$

The following result and its proof have been shown by Toda (private communication).

Theorem 1.9. *Let $n \geq 2$. Then, $[\iota_{2n}, [\iota_{2n}, \alpha_1(2n)]] \neq 0$ if and only if $n \neq 2$ and $2n \equiv 1 \pmod 3$.*

Proof. First of all, observe that using [25, Corollary 2.4] and the proof of [32, Corollary (5.9)], the formula

$$[[\alpha, \beta], \gamma] \in E\pi_{6n-2}(\mathbb{S}^m) \text{ for } \alpha, \beta, \gamma \in \pi_{2n}(\mathbb{S}^m) \tag{1.13}$$

holds. By (1.4), (1.6) and (1.13), we obtain

$$[\iota_{2n}, [\iota_{2n}, \iota_{2n}]] = J\Delta[\iota_{2n}, \iota_{2n}] \in E\pi_{6n-3}(\mathbb{S}^{2n-1};3). \tag{1.14}$$

By (1.11) and (1.12), $[\iota_{2n}, [\iota_{2n}, \iota_{2n}]] = \pm EP(i(2n-1))$. By the naturality [87, (2.1.5)], we obtain $[\iota_{2n}, [\iota_{2n}, \alpha_1(2n)]] = \pm EP(a_1(2n-1))$. By [87, (4.15), Proposition 4.4], $(n+1)a_1(2n-1) = HP(i(2n+1))$. So, $P(a_1(2n-1)) = \pm PHP(i(2n+1)) = 0$ if $2n \not\equiv 1 \pmod 3$. For the case $n = 2$, the assertion is trivial.

Next, assume that $n \neq 2$ and $2n \equiv 1 \pmod 3$. Then, by [87, Theorem 10.3], there exists an element $v \in \pi_{6n-2}(\mathbb{S}^{2n-3})$ satisfying $H(v) = b_1(2n-5)$ and $E^2 v = P(a_1(2n-1))$. Furthermore, by [87, Proposition 5.3(ii)], we obtain $P(a_2(2n-3)) = 3v$. Hence, by the (mod 3) EHP-sequence (1.11), we have $P(a_1(2n-1)) \neq 0$. This implies the sufficient condition and completes the proof. □

We show

Proposition 1.10. 1. *Let* $3 \leq n \leq 27$. *Then,* $G_{4n+2}(\mathbb{S}^{2n};3) = 0$ *if* $n = 5, 8, 11, 14, 17, 20, 23, 26$ *and* $G_{4n+2}(\mathbb{S}^{2n};3) = \{[\iota_{2n}, \alpha_1(2n)]\} \cong \mathbb{Z}_3$ *otherwise.*
2. *Let* $3 \leq n \leq 9$. *Then,* $G_{6n-2}(\mathbb{S}^{2n};3) = \{[\iota_{2n}, [\iota_{2n}, \iota_{2n}]]\} \cong \mathbb{Z}_3$ *for* $n = 3, 5, 9$,
$G_{22}(\mathbb{S}^8;3) = \{[\iota_8, [\iota_8, \iota_8]], [\iota_8, \alpha_2(8)]\} \cong (\mathbb{Z}_3)^2$,
$G_{34}(\mathbb{S}^{12};3) = \{[\iota_{12}, [\iota_{12}, \iota_{12}]], [\iota_{12}, \alpha_3'(12)]\} \cong \mathbb{Z}_3 \oplus \mathbb{Z}_9$,
$G_{40}(\mathbb{S}^{14};3) = \{[\iota_{14}, [\iota_{14}, \iota_{14}]], [\iota_{14}, \alpha_1(14)\beta_1(17)]\} \cong (\mathbb{Z}_3)^2$, *and*
$G_{46}(\mathbb{S}^{16};3) = \{[\iota_{16}, [\iota_{16}, \iota_{16}]], [\iota_{16}, \alpha_4(16)]\} \cong (\mathbb{Z}_3)^2$.

Proof. Notice that $G_{6n-2}(\mathbb{S}^{2n}) \ni [\iota_{2n}, [\iota_{2n}, \iota_{2n}]]$ by Lemma 1.2(7).

The assertion is obtained from [87, pp. 60–61: Table], (1.7), (1.4), Theorem 1.9. We determine $\pi_{38}(\mathbb{S}^{18};3)$ and $\pi_{34}(\mathbb{S}^{12};3)$. The rest is similar.

1. By [87, pp. 60–61: Table], $\pi_{n+20}(\mathbb{S}^n;3) = \{\beta_1^2(n)\} \cong \mathbb{Z}_3$ for $n \geq 5$. Then, by (1.7), $\pi_{38}(\mathbb{S}^{18};3) = \{\beta_1^2(18), [\iota_{18}, \alpha_1(18)]\} \cong (\mathbb{Z}_3)^2$. Again, by (1.7), we get $[\iota_{18}, \beta_1^2(18)] \neq 0$. Hence, by Theorem 1.9, $G_{38}(\mathbb{S}^{18};3) = \{[\iota_{18}, \alpha_1(18)]\} \cong \mathbb{Z}_3$.
2. By (1.7), $\pi_{34}(\mathbb{S}^{12};3) = E\pi_{23}(\mathbb{S}^{11};3) \oplus \{[\iota_{12}, \iota_{12}] \circ \alpha_3'(23)\}$. By [87, pp. 60–61: Table] and (1.14), $[\iota_{12}, [\iota_{12}, \iota_{12}]] \in E^3\pi_{31}(\mathbb{S}^9;3)$ and so, $[\iota_{12}, [\iota_{12}, \alpha_3'(12)]] \in E^3\pi_{42}(\mathbb{S}^9;3)$. Moreover, $\pi_{42}(\mathbb{S}^9;3) \cong \mathbb{Z}_3$ and $E^4 : \pi_{42}(\mathbb{S}^9;3) \to \pi_{45}(\mathbb{S}^{13};3) \cong \mathbb{Z}_9$ is injective. This implies $[\iota_{12}, [\iota_{12}, \alpha_3'(12)]] = 0$ and hence the group $G_{34}(\mathbb{S}^{12};3)$ follows.

□

Remark 1.11. In virtue of (1.13) and Lemma 1.1.(2);(6), $[\iota_{2n}, [\iota_{2n}, [\iota_{2n}, \iota_{2n}]]] = [\iota_{2n}, \iota_{2n}] \circ E^{2n-1}[\iota_{2n}, [\iota_{2n}, \iota_{2n}]] = 0$.

Write $[X, Y]$ for the set of homotopy classes of maps from X to Y. In the sequel, we need the following fact proved in [55, Theorem 5.2]:

Proposition 1.12. *Assume that* $\alpha \circ E\beta = \beta \circ \gamma = 0$ *for* $\alpha \in [E^{n+1}X, W]$, $\beta \in [E^nY, E^nX]$ *and* $\gamma \in [E^nZ, E^nY]$, *where* W, X, Y, Z *are pointed CW-complexes. Then, we have*

$$\Delta_{\mathbb{F}}\{\alpha, E^n\beta, E^n\gamma\}_n \subseteq \{\Delta_{\mathbb{F}}(\alpha), E^{n-1}\beta, E^{n-1}\gamma\}_{n-1}.$$

1.2 Gottlieb Groups of Spheres with Stems for $k \leq 7$

According to [28, Sect. 4], [26, Theorem], [38, Theorem 1.3], [43, Theorem 1.1.2(a)], [68, Theorem], [84, 2.7, 2.8 and 2.13], and [85, Chaps. VII, X, XII], we know the following results:

$$\sharp[\iota_n, \eta_n] = \begin{cases} 1 \text{ for } n = 2, 6 \text{ or } n \equiv 3 \pmod 4, \\ 2 \text{ for otherwise;} \end{cases} \tag{1.15}$$

and

$$\sharp[\iota_n, \eta_n^2] = \begin{cases} 1 \text{ for } n \equiv 2, 3 \pmod 4, \\ 2 \text{ for otherwise.} \end{cases} \tag{1.16}$$

Hence, Lemma 1.1 completely determines $G_{n+k}(\mathbb{S}^n)$ for $k = 1, 2$.

Set ν' and $\sigma''', \sigma'', \sigma'$ for the generators of $\pi_6^3 \cong \mathbb{Z}_4$ and $\pi_{n+7}^n \cong \mathbb{Z}_{2^{n-4}}$ with $5 \leq n \leq 7$, respectively. Next, write $\eta_2 \in \pi_3^2$, $\nu_4 \in \pi_7^4$ and $\sigma_8 \in \pi_{15}^8$ for the elements defined in [85, Chap. V]. Then, $\eta_n = E^{n-2}\eta_2$, $\nu_n = E^{n-4}\nu_4$, and $\sigma_n = E^{n-8}\sigma_8$ generate $\pi_{n+1}^n = \mathbb{Z}_2\{\eta_n\}$ for $n \geq 3$, $\pi_{n+3}^n = \mathbb{Z}_8\{\nu_n\}$ for $n \geq 5$ and $\pi_{n+7}^n = \mathbb{Z}_{16}\{\sigma_n\}$ for $n \geq 9$, respectively.

Notice, by means of Toda brackets, the relations:

$$\begin{aligned} &(1)\ \eta_n^2 = \{2\iota_n, \eta_n, 2\iota_{n+1}\} \text{ for } n \geq 3, \\ &(2)\ \pm\nu' = \{\eta_3, 2\iota_4, \eta_4\}_1, \\ &(3)\ \sigma''' = \{\nu_5, 8\iota_8, \nu_8\}_t \text{ for } 0 \leq t \leq 3 \text{ [85, Lemma 5.13].} \end{aligned} \tag{1.17}$$

Further, recall that

$$\eta_3^3 = 2\nu' \text{ [85, (5.3)]} \tag{1.18}$$

and

$$\pi_{n+3}^n = \{\nu_n\} \cong \mathbb{Z}_8 \text{ for } n \geq 5 \text{ [85, Proposition 5.6].} \tag{1.19}$$

We know that $\pi_{n+7}(\mathbb{S}^n; 5) = \mathbb{Z}_5\{\alpha_1'(n)\}$ for $n \geq 3$. Set

$$\nu_n^+ = \nu_n + \alpha_1(n) \ \text{ for } \ n \geq 4$$

and

$$\sigma_n^+ = \sigma_n + \alpha_2(n) + \alpha_{1,5}(n) \ \text{ for } \ n \geq 8$$

which stand for generators of $\pi_{n+3}(\mathbb{S}^n) \cong \mathbb{Z}_{24}$ for $n \geq 5$ and $\pi_{n+7}(\mathbb{S}^n) \cong \mathbb{Z}_{240}$ for $n \geq 9$, respectively.

We determine ν_5^+ and σ_9^+ satisfying the equations $E^2\nu'^+ = 2\nu_5^+$ and $E^2\sigma'^+ = 2\sigma_9^+$. To aim that, we set $\nu'^+ = \nu' + x\alpha_1(3)$ and $\sigma'^+ = \sigma' + y\alpha_1(7) + z\alpha_{1,5}(7)$. Then, $x \equiv 2 \pmod 3$ and we get

$$\nu'^+ = \nu' - \alpha_1(3)$$

which generates the J-image $J\pi_3(SO(3)) = \pi_6(\mathbb{S}^3) \cong \mathbb{Z}_{12}$.

Further, we obtain $y \equiv 2 \pmod 3$ and $z \equiv 2 \pmod 5$ which lead to

$$\sigma'^+ = \sigma' + 2\alpha_2(7) + 2\alpha_{1,5}(7)$$

which generates $\pi_{14}(\mathbb{S}^5) \cong \mathbb{Z}_{120}$.

We recall that

$$\pm [\iota_4, \iota_4] = 2\nu_4 - E\nu' = 2\nu_4^+ - E\nu'^+ \text{ [85, (5.8)]}, \tag{1.20}$$

$$\nu_5\eta_8 = [\iota_5, \iota_5] \text{ [85, (5.10)]}, \tag{1.21}$$

and

$$\pm [\iota_8, \iota_8] = 2\sigma_8 - E\sigma' = 2\sigma_8^+ - E\sigma'^+ \text{ [85, (5.16)]}. \tag{1.22}$$

Then, the relations (1.20), (1.21), and (1.33) yield

$$\eta_5\nu_6 = 0 \quad \text{and} \quad \nu_6\eta_9 = 0. \tag{1.23}$$

This leads to the relations:

$$\eta_n\nu_{n+1} = 0 \text{ for } n \geq 5 \text{ and } \nu_n\eta_{n+3} = 0 \text{ for } n \geq 6.$$

Further, we obtain

$$[\iota_5, \nu_5] = 0. \tag{1.24}$$

In view of the relations above, we get:

$$\begin{aligned} E^{n-3}\nu' &= 2\nu_n, \\ \eta_n^3 &= 4\nu_n, \quad \text{for } n \geq \\ E^{n-3}\nu'^+ &= 2\nu_n^+5 \end{aligned} \tag{1.25}$$

and

$$\begin{aligned} E^{n-7}\sigma' &= 2\sigma_n, \\ E^{n-7}\sigma'^+ &= 2\sigma_n^+ \quad \text{for } n \geq 9. \end{aligned} \tag{1.26}$$

In the sequel we need:

$$\begin{aligned} &(1)\ \pi_k(\mathbb{S}^2) = \eta_{2*}\pi_k(\mathbb{S}^3), \\ &(2)\ \pi_k(\mathbb{S}^4) = \nu^+_{4*}\pi_k(\mathbb{S}^7) \oplus E\pi_{k-1}(\mathbb{S}^3), \\ &(3)\ \pi_k(\mathbb{S}^8) = \sigma^+_{8*}\pi_k(\mathbb{S}^{15}) \oplus E\pi_{k-1}(\mathbb{S}^7), \\ &(4)\ [\iota_4, \nu_4] = 2\nu_4^2 \text{ Proposition 1.6.} \end{aligned} \tag{1.27}$$

Next, $\pi_7(\mathbb{S}^4) = \mathbb{Z}\{\nu_4\} \oplus \mathbb{Z}_{12}\{E\nu'^+\}$ and $\pi_{n+3}(\mathbb{S}^n) = \mathbb{Z}_{24}\{\nu_n^+\}$ for $n \geq 5$. In light of Lemma 1.2(2) and the relation

$$\nu'\nu_6 = 0 \text{ [85, Proposition 5.11]} \tag{1.28}$$

we obtain

$$[\iota_4, E\nu'] = [\iota_4, \iota_4] \circ E(\iota_3 \wedge \nu') = (2\nu_4 - E\nu') \circ 2\nu_7 = 4\nu_4^2. \tag{1.29}$$

Notice that $\nu' E^3\alpha \neq 0$ for $\alpha = \eta_3, \eta_3^2$ [85, Propositions 5.8, 5.9]. Further, (1.7) and (1.29) lead to

$$\sharp[\iota_4, E\nu'^+] = 6. \tag{1.30}$$

The equation (1.18) yields $2E\nu' = \eta_4^3 \in G_7(\mathbb{S}^4)$. Consequently, by Corollary 1.3(1) and Proposition 1.7,

$$G_7(\mathbb{S}^4) = \{3[\iota_4, \iota_4], 2E\nu'\} \cong 3\mathbb{Z} \oplus \mathbb{Z}_2.$$

In light of [38, (13)], [43, Theorem 1.1.2(b)], [68, Theorem], [84, 2.10 and 2.14], [85, Chaps. VII, X and XII], Corollary 1.3(3), and Proposition 1.7, we know the following:

$$\sharp[\iota_n, \nu_n] = \begin{cases} 1 & \text{for } n \equiv 7 \pmod 8,\ n = 2^i - 3 \geq 5; \\ 2 & \text{for } n \equiv 1, 3, 5 \pmod 8 \text{ with } n \geq 9 \text{ and } n \neq 2^i - 3; \\ 4 & \text{for } n \equiv 2 \pmod 4 \text{ with } n \geq 6 \text{ or } n = 4, 12; \\ 8 & \text{for } n \equiv 0 \pmod 4 \text{ with } n \geq 8 \text{ and } n \neq 12 \end{cases} \tag{1.31}$$

which yields

$$\sharp[\iota_n, \nu_n^+] = \begin{cases} 1, & \text{if } n \equiv 7 \pmod 8 \text{ or } n = 2^i - 3 \text{ for } i \geq 3; \\ 2, & \text{if } n \equiv 1, 3, 5 \pmod 8,\ n \geq 9 \text{ and } n \neq 2^i - 3; \\ 12, & \text{if } n \equiv 2 \pmod 4 \text{ and } n \geq 6 \text{ or } n = 4, 12; \\ 24, & \text{if } n \equiv 0 \pmod 4 \text{ and } n \geq 8 \text{ unless } n = 12. \end{cases} \tag{1.32}$$

Thus, (1.19) and Lemma 1.1 lead to a complete description of $G_{n+3}(\mathbb{S}^n)$.

By (1.27)(4), we get $[\iota_4, \nu_4\eta_7] = (2\nu_4^2)\eta_{10} = 0$ and (1.29) leads to $[\iota_4, (E\nu')\eta_7] = (4\nu_4^2)\eta_{10} = 0$. Next, (1.21) yields $[\iota_5, \nu_5\eta_8] = [\iota_5, [\iota_5, \iota_5]] = 0$. Recall that

$$\eta_3\nu_4 = \nu'\eta_6 \text{ [85, (5.9)].} \tag{1.33}$$

We know that $\pi_{n+4}(\mathbb{S}^n) = 0$ for $n \geq 6$ [85, Proposition 5.8], $\pi_{11}(\mathbb{S}^6) = \{[\iota_6, \iota_6]\} \cong \mathbb{Z}$ and $\pi_{n+5}(\mathbb{S}^n) = 0$ for $n \geq 7$ [85, Proposition 5.9]. Hence, by the group structures of $\pi_{n+k}(\mathbb{S}^n)$ for $k = 4, 5$, we get

Proposition 1.13. *$G_{n+4}(\mathbb{S}^n) = \pi_{n+4}(\mathbb{S}^n)$; $G_{n+5}(\mathbb{S}^n) = \pi_{n+5}(\mathbb{S}^n)$ unless $n = 6$ and $G_{11}(\mathbb{S}^6) = 3\pi_{11}(\mathbb{S}^6) \cong 3\mathbb{Z}$.*

Now, recall

$$\pi_{n+6}^n = \mathbb{Z}_2\{\nu_n^2\} \text{ for } n \geq 5 \text{ [85, Proposition 5.11]} \tag{1.34}$$

In the next two sections, we will prove the following result partially extending that of [38, Theorem 1.3].

Theorem 1.14. *$[\iota_n, \nu_n^2] = 0$ if and only if $n \equiv 4, 5, 7 \pmod 8$ or $n = 2^i - 5$ for $i \geq 4$.*

We recall that $\pi_{10}(\mathbb{S}^4) = \mathbb{Z}_8\{\nu_4^2\} \oplus \mathbb{Z}_3\{\alpha_1(4)\alpha_1(7)\} \oplus \mathbb{Z}_3\{\nu_4\alpha_1(7)\}$.

By (1.8), we get that $[\iota_4, \alpha_1(4)\alpha_1(7)] = [\iota_4, \iota_4]\alpha_1(7)\alpha_1(10) = 0$ and (1.27)(4) leads to $[\iota_4, \nu_4\alpha_1(7)] = (2\nu_4^2)\alpha_1(10) = 0$. Recall also that $\pi_{12}^5 = \{\sigma'''\} \cong \mathbb{Z}_2$, $\pi_{13}^6 = \{\sigma''\} \cong \mathbb{Z}_4$ and $\pi_{14}^7 = \{\sigma'\} \cong \mathbb{Z}_8$. Then, (1.22) and [85, Lemma 5.14] lead to:

$$\begin{aligned} &(1)\ E\sigma'' = 2\sigma',\\ &(2)\ E\sigma''' = 2\sigma'',\\ &(3)\ E^{n-7}\sigma' = 2\sigma_n,\\ &(4)\ E^{n-6}\sigma'' = 4\sigma_n,\\ &(5)\ E^{n-5}\sigma''' = 8\sigma_n \text{ for } n \geq 9 \end{aligned} \tag{1.35}$$

and

$$\eta_4\sigma''' = 0 \text{ [85, (7.4)].} \tag{1.36}$$

By (1.35) and Proposition 1.6, we obtain

$$[\iota_5, \sigma'''] = [\iota_5, \iota_5] \circ E^4\sigma''' = 0, \tag{1.37}$$

$$[\iota_6, \sigma''] = [\iota_6, \iota_6] \circ E^5\sigma'' = 4([\iota_6, \iota_6] \circ \sigma_{11})$$

and

$$\sharp[\iota_6, \sigma''] = 4. \tag{1.38}$$

Further,

$$\nu_6\mu_9 = 8P(\sigma_{13}) = 2[\iota_6, \sigma''] = [\iota_6, E\sigma'''] \text{ [85, Lemma 5.14 and (7.25)].} \quad (1.39)$$

Denote by $M^n = E^{n-2}\mathbb{R}P^2$ the Moore space of type $(\mathbb{Z}_2, n-1)$ for $n \geq 3$.

Given the inclusion $i_2 : \mathbb{S}^1 \hookrightarrow \mathbb{R}P^2$ and collapsing $p_2 : \mathbb{R}P^2 \to \mathbb{R}P^2/\mathbb{S}^1 = \mathbb{S}^2$ maps, respectively, we write $i_n = E^{n-2}i_2 : \mathbb{S}^{n-1} \to M^n$ and $p_n = E^{n-2}p_2 : M^n \to \mathbb{S}^n$ for $n \geq 2$. Certainly, $M^n = \mathbb{S}^{n-1} \cup_{2\iota_{n-1}} e^n$ and there is the cofiber sequence

$$(\mathcal{CS}_n): \quad \mathbb{S}^{n-1} \xrightarrow{i_n} M^n \xrightarrow{p_n} \mathbb{S}^n \xrightarrow{2\iota_n} \mathbb{S}^n.$$

Notice that $\pi_{n-1}(M^n) = \mathbb{Z}_2\{i_n\}$ for $n \geq 3$. Further, using [63], we state:

Lemma 1.15. *$\pi_3(M^3) = \mathbb{Z}_4\{i_3\eta_2\}$ and $\pi_n(M^n) = \mathbb{Z}_2\{i_n\eta_{n-1}\}$ for $n \geq 4$.*

In view of [86, p. 307, Corollary], it holds

$$2\iota_{M^n} = i_n\eta_{n-1}p_n, \text{ if } n \geq 4.$$

Next, the short exact sequence

$$0 \longrightarrow \text{Ext}(\mathbb{Z}_2, \pi_3(M^3)) \longrightarrow [M^3, M^3] \longrightarrow \text{Hom}(\mathbb{Z}_2, \pi_2(M^3)) \longrightarrow 0$$

determined by the universal coefficient theorem for homotopy [27, p. 30] and Lemma 1.15(2) lead to $\sharp[M^3, M^3] = 4$. Since the suspension map $E : [M^3, M^3] \to [M^4, M^4]$ is an epimorphism, the above yields that this map is an isomorphism. Thus, we derive that

$$2\iota_{M^n} = i_n\eta_{n-1}p_n, \text{ if } n \geq 3 \quad (1.40)$$

which leads to

$$\sharp\iota_{M^n} = 4 \text{ for } n \geq 3. \quad (1.41)$$

Further, by [63, Lemma 1.1], it holds $[M^n, M^n] = \mathbb{Z}_4\{\iota_{M^n}\}$ for $n \geq 3$.

Next, we recall from [85, pp. 95–96] the construction of the element $\kappa_7 \in \pi_{21}(\mathbb{S}^7)$. It is a representative of a Toda bracket

$$\{\nu_7, E\alpha, E^2\beta\}_1,$$

where $\alpha = \bar{\eta}_9 \in [M^{11}, \mathbb{S}^9]$ is an extension of η_9 and $\beta = \tilde{\bar{\nu}}_9 \in \pi_{18}(M^{10})$ is a coextension of $\bar{\nu}_9$ satisfying $\alpha \circ E\beta = 0$. Furthermore, $\kappa_n = E^{n-7}\kappa_7$ for $n \geq 7$ and set $\tilde{\bar{\nu}}_n = E^{n-9}\tilde{\bar{\nu}}_9$ for $n \geq 9$. Then, we can take

$$\kappa_n \in \{\nu_n, \bar{\eta}_{n+3}, \tilde{\bar{\nu}}_{n+4}\} \text{ for } n \geq 7.$$

In $\pi_{22}^{8} = \mathbb{Z}_{16}\{\sigma_8^2\} \oplus \mathbb{Z}_8\{(E\sigma')\sigma_{15}\} \oplus \mathbb{Z}_4\{\kappa_8\}$, we have $[\iota_8, E\sigma'] = 2[\iota_8, \iota_8]\sigma_{15} = \pm 2(2\sigma_8^2 - (E\sigma')\sigma_{15})$ and

$$\sharp[\iota_8, E\sigma'] = 4. \tag{1.42}$$

Further, in view of Proposition 1.6 and (1.22), we obtain $[\iota_8, \sigma_8] = [\iota_8, \iota_8] \circ \sigma_{15} = \pm(2\sigma_8^2 - (E\sigma')\sigma_{15})$. Thus, (1.34), Corollary 1.3, Proposition 1.7, and Theorem 1.14 yield

Proposition 1.16. *$G_{n+6}(\mathbb{S}^n) = \pi_{n+6}(\mathbb{S}^n)$ if $n \equiv 4, 5, 7 \pmod 8$ or $n = 2^i - 5$ and $G_{n+6}(\mathbb{S}^n) = 0$ otherwise.*

Furthermore, $G_{n+7}(\mathbb{S}^n) = 0$ if $n = 4, 6$, $G_{12}(\mathbb{S}^5) = \pi_{12}(\mathbb{S}^5)$ and $G_{15}(\mathbb{S}^8) = \{3[\iota_8, \iota_8], 4E\sigma'\} \cong 3\mathbb{Z} \oplus \mathbb{Z}_2$.

By [1] and [72, Proposition 4.5], there exists an element $\gamma \in \pi_{2n-8}^{n-7}$ satisfying

$$[\iota_n, \iota_n] = E^7\gamma \text{ and } H\gamma = \sigma_{2n-15}, \text{ if } n \equiv 7 \pmod{16} \text{ and } n \geq 23. \tag{1.43}$$

Concerning (1.43), we obtain

Theorem 1.17 (Mahowald (private communication)).

1. $[\iota_n, \sigma_n] \neq 0$, *if $n \equiv 7 \pmod{16}$ and $n \geq 23$.*
2. *It desuspends seven dimensions whose Hopf invariant is σ_{2n-15}^2.*

In virtue of Theorem 1.43(2), Theorem 1.17(1) is obtained and this will be proved in Sect. 1.6. By [85, (10.18), Theorem 10.5],

$$[\iota_9, \sigma_9] = \sigma_9(\bar{\nu}_{16} + \varepsilon_{16}) \neq 0 \tag{1.44}$$

and

$$\sigma_{11}\bar{\nu}_{18} = \sigma_{11}\varepsilon_{18} = 0. \tag{1.45}$$

In view of [85, Theorem 12.16], $\sharp[\iota_{10}, \sigma_{10}] = 16$ and, by [85, Lemma 12.14],

$$[\iota_{11}, \sigma_{11}] = 0. \tag{1.46}$$

We know that $\sharp[\iota_{12}, \sigma_{12}] = 16$ [85, Lemma 12.19, Theorem 12.22], and $[\iota_{13}, \sigma_{13}] \neq 0$ [85, p. 166]. We also know that $\sharp[\iota_{14}, \sigma_{14}] = 16$ [53, p. 52], $[\iota_{15}, \sigma_{15}] = 0$ [49, Lemma 6.2], $\sharp[\iota_{16}, \sigma_{16}] = 16$ [49, p. 323], $[\iota_{17}, \sigma_{17}] \neq 0$ [52, p. 27], and $\sharp[\iota_{18}, \sigma_{18}] = 16$ [52, (5.36)]. By [73, p. 72: (7.23)], $[\iota_{19}, \sigma_{19}] \neq 0$. By [73, p. 142, Theorem 3(b)], $\sharp[\iota_{20}, \sigma_{20}] = 16$.

Let $\mathbb{R}$ and $\mathbb{C}$ be the fields of real and complex numbers, respectively, and $\mathbb{H}$ the skew $\mathbb{R}$-algebra of quaternions and put $d = \dim_{\mathbb{R}}\mathbb{F}$ for $\mathbb{F} = \mathbb{R}, \mathbb{C}$ or $\mathbb{H}$, respectively. Denote by $r = r_n : U(n) \longrightarrow SO(2n)$ and $c = c_n : Sp(n) \longrightarrow SU(2n)$ the canonical map. Let $J = J_{\mathbb{R}} : \pi_k(SO(n)) \longrightarrow \pi_{n+k}(\mathbb{S}^n)$ be the J-homomorphism, $J_{\mathbb{F}} : \pi_k(SO_{\mathbb{F}}(n)) \longrightarrow \pi_{k+dn}(\mathbb{S}^{dn})$ the complex or symplectic J-homomorphism defined as follows: $J_{\mathbb{C}} = J \circ r_* : \pi_k(SU(n)) \longrightarrow \pi_{2n+k}(\mathbb{S}^{2n})$ and $J_{\mathbb{H}} = J \circ r_* \circ c_* : \pi_k(Sp(n)) \longrightarrow \pi_{4n+k}(\mathbb{S}^{4n})$.

Write $\omega_{n,\mathbb{F}} \in \pi_{d(n+1)-2}(O_{\mathbb{F}}(n))$ for the characteristic map of the canonical bundle $O_{\mathbb{F}}(n+1) \xrightarrow{O_{\mathbb{F}}(n)} \mathbb{S}^{d(n+1)-1}$. We note that $\omega_{n,\mathbb{F}} = \Delta_{\mathbb{F}}\iota_{(n+1)d-1}$ and $\omega_{1,\mathbb{H}} = \nu'^{+}$.

As it is well known (see, e.g., [82, Part II])

$$i_{2n+1,\mathbb{R}} r_n \omega_{n,\mathbb{C}} = \omega_{2n+1,\mathbb{R}} \text{ and } i_{2n+1,\mathbb{C}} c_n \omega_{n,\mathbb{H}} = \omega_{2n+1,\mathbb{C}}. \tag{1.47}$$

Further,

$$(1)\ i_{2n+1,\mathbb{R}*} r_{n*} \Delta_{\mathbb{C}} = \Delta \text{ and } (2)\ i_{2n+1,\mathbb{C}*} c_{n*} \Delta_{\mathbb{H}} = \Delta_{\mathbb{C}}. \tag{1.48}$$

Then, in view of $r_{2n+1} i_{2n+1,\mathbb{C}} = i_{4n+2,\mathbb{R}} i_{4n+1,\mathbb{R}} r_{2n}$ and [85, Corollary 11.2], we see that

$$E^{d-1} \circ J_{\mathbb{F}} \circ \Delta_{\mathbb{F}} = -J \circ \Delta \text{ for } \mathbb{F} = \mathbb{C}, \mathbb{H} \text{ and } \Delta = \Delta_{\mathbb{R}}. \tag{1.49}$$

Combining the results of [43, 45, 85], $J(\Delta(E\sigma')) = \pm[\iota_8, E\sigma']$, Corollary 1.3(3), Proposition 1.7, and Theorem 1.17,

$$\sharp[\iota_n, \sigma_n] = \begin{cases} 8 & \text{for } n = 8; \\ 1 & \text{for } n = 11,\ n \equiv 15 \pmod{16}; \\ 2 & \text{for } n \equiv 1 \pmod 2 \text{ with } n \geq 9 \text{ and } n \neq 11,\ n \not\equiv 15 \pmod{16}; \\ 16 & \text{for } n \equiv 0 \pmod 2 \text{ with } n \geq 10 \end{cases} \tag{1.50}$$

which yields

$$\sharp[\iota_n, \sigma_n^{+}] = \begin{cases} 1, & \text{if } n = 11 \text{ or } n \equiv 15 \pmod{16}; \\ 2, & \text{if } n \text{ is odd and } n \geq 9 \text{ unless } n = 11 \text{ and } n \equiv 15 \pmod{16}; \\ 120, & \text{if } n = 8; \\ 240, & \text{if } n \text{ is even and } n \geq 10. \end{cases} \tag{1.51}$$

Whence, by means of Lemma 1.1, the groups $G_{n+7}(\mathbb{S}^n)$ for $n \geq 9$ have been fully described as well.

1.3 Proof of Theorem 1.14, Part I

Since $SO(n) \cong SO(n-1) \times \mathbb{S}^{n-1}$ for $n = 4, 8$, we get that

$$\Delta\pi_k(\mathbb{S}^n) = 0, \text{ if } n = 3, 7. \tag{1.52}$$

By the exact sequence $(\mathcal{SO}^n_n)$ and the fact that $\pi_n(SO(n)) \cong \mathbb{Z}$ for $n \equiv 3 \pmod 4$ [37], we have

$$\Delta\eta_n = 0, \text{ if } n \equiv 3 \pmod 4. \tag{1.53}$$

We recall the formula [37, Lemma 1]

$$\Delta(\alpha \circ E\beta) = \Delta\alpha \circ \beta. \tag{1.54}$$

By (1.53) and (1.54),

$$\Delta(\eta_n^2) = 0, \text{ if } n \equiv 3 \pmod 4. \tag{1.55}$$

Given $\alpha \in \pi_{n+k}(\mathbb{S}^n)$ and $\beta \in \pi_{n+k}(SO(n+1))$ satisfying $p_{n+1}(\mathbb{R})\beta = \alpha$, the element β is called a *lift* of α and we put

$$\beta = [\alpha]. \tag{1.56}$$

For $m \leq n-1$, set $i_{m,n} = i_n(\mathbb{R}) \circ \cdots \circ i_{m+1}(\mathbb{R})$. We write $[\alpha]_n = i_{m,n*}[\alpha] \in \pi_k(SO(n))$, where $[\alpha] \in \pi_k(SO(m))$ is a lift of $\alpha \in \pi_k(\mathbb{S}^{m-1})$.

By [85, Chap. VI, i)], the Toda bracket $\{\eta_3, E\nu', \nu_7\}_1$ consists a single $\varepsilon_3 \in \pi_{11}^3$. Further, $\varepsilon_n = E^{n-3}\varepsilon_3$ for $n \geq 3$ and, in view of [85, (6.1)], the following Toda brackets contain ε_n:

$$\begin{aligned}
&(1)\ \{\eta_n, 2\nu_{n+1}, \nu_{n+4}\}_t \text{ for } n \geq 4 \text{ and } 0 \leq t \leq n-2;\\
&(2)\ \{\eta_n, \nu_{n+1}, 2\nu_{n+4}\}_t,\ \{\eta_n, 2\iota_{n+1}, \nu_{n+1}^2\}_t \text{ and}\\
&(3)\ \{\eta_n, \nu_{n+1}^2, 2\iota_{n+7}\}_t \text{ for } n \geq 5, \text{ and } 0 \leq t \leq n-5.
\end{aligned} \tag{1.57}$$

Notice that the indeterminacy of those Toda brackets are:

$$\begin{aligned}
&\pi_{n+5}^n \circ \nu_{n+5} + \eta_n \circ E^t\pi_{n-t+8}^{n-t+1},\\
&\pi_{n+5}^n(2\nu_{n+5}) + \eta_n \circ E^t\pi_{n-t+8}^{n-t+1},\\
&\pi_{n+2}^n \circ \nu_{n+2}^2 + \eta_n \circ E^t\pi_{n-t+8}^{n-t+1} \text{ and}\\
&\pi_{n+8}^n(2\iota_{n+8}) + \eta_n \circ E^t\pi_{n-t+8}^{n-t+1}
\end{aligned}$$

respectively.

We recall that $\pi_{n+8}(\mathbb{S}^n) = \{\varepsilon_n\} \cong \mathbb{Z}_2$ for $3 \leq n \leq 5$. Further, in view of the proof of [85, (6.1)], it holds

$$2\iota_6 \circ \sigma'' = E\sigma'''. \tag{1.58}$$

Then, (1.58) yields $\eta_5 \circ E\pi_{12}^5 = \{\eta_5 \circ E\sigma'''\} = \{\eta_5 \circ 2\iota_6 \circ \sigma''\} = 0$. Next, $\pi_{12}^5 = \{\nu_5\eta_8^2\}$ and (1.23) lead to $\pi_{12}^5 \circ \nu_{10} = \pi_{12}^5 \circ (2\nu_{10}) = \pi_{12}^5 \circ \nu_7^2 = \pi_{12}^5 \circ (2\iota_{10}) = 0$. Consequently,

$$\varepsilon_5 = \{\eta_5, 2\nu_6, \nu_9\}_1 = \{\eta_5, \nu_6, 2\nu_6\}_1 = \{\eta_5, 2\iota_6, \nu_6^2\}_1 = \{\eta_5, \nu_5^2, 2\iota_{12}\}. \tag{1.59}$$

Because $\pi_{n+5}^n = 0$ for $n \geq 7$ ([85, Proposition 5.9]) and $2\pi_{n+8}^n = 0$ for $n \geq 7$ ([85, Theorem 7.1]), the relation (1.23) shows all those Toda brackets are equal for $n \geq 7$ and the indicated t.

By [85, Proposition 1.9], there exists an extension $\beta : EK \to \mathbb{S}^3$ of ν' such that $\eta_3 \circ E\beta = 0$, where $K = e^6 \cup_{8\iota_5} \mathbb{S}^5$. Next, consider a coextension $\gamma : \mathbb{S}^9 \to K$ of ν_5, then [85, Proposition 1.7] implies that $\beta \circ E\gamma \in \{\nu', 8\iota_8, \nu_6\}_1 \subseteq \pi_{10}^3 = 0$. Finally, choose $\mu_3 \in \{\eta_3, E\beta, E^2\gamma\}_1 \in \pi_{12}(\mathbb{S}^3)/\{\eta_3\varepsilon_4\}$. Then,

$$\mu_n = E^{n-3}\mu_3 \in \{\eta_n, 2\iota_{n+1}, E^{n-4}\sigma'''\}_{n-4} + \{\nu_n^3\} \text{ for } n \geq 4 \text{ [85, Lemma 6.5]}, \tag{1.60}$$

$$\varepsilon_n\eta_{n+8} = \eta_n\varepsilon_{n+1} \text{ for } n \geq 3 \text{ [85, (7.5)]}, \tag{1.61}$$

$$\mu_n\eta_{n+9} = \eta_n\mu_{n+1} \text{ for } n \geq 3 \text{ [75, Proposition 2.2(2)]}, \tag{1.62}$$

and

$$\mu_n\sigma_{n+9} = \sigma_n\mu_{n+7} \text{ for } n \geq 11 \text{ [85, Proposition 3.1]}. \tag{1.63}$$

In view of [85, Lemma 5.2], we have an element $\mu' \in \{\eta_3, 2\iota_4, \mu_4\}_1$ such that $H(\mu') = \mu_5$ and $2\mu' = \eta_3^2\mu_5$.

Further, we follow [85, Chap. VI, v)] to choose $\zeta_5 \in \{\nu_5, 8\iota_8, E\sigma'\}_1 \in \pi_{16}^5/(\nu_5 \circ E\pi_{15}^7)$ and we write:

$$\zeta_n^+ = \zeta_n + \alpha_3'(n) + \alpha_{1,7}(n) \quad \text{for} \quad n \geq 5.$$

Then

$$\left.\begin{array}{l}(1)\ 2\zeta_5 = \pm E^2\mu', \\ (2)\ 4\zeta_n = \eta_n^2\mu_{n+2} \text{ for } n \geq 5\end{array}\right. \text{[85, (7.14)]} \tag{1.64}$$

and

$$\sigma'\zeta_{14} = x\zeta_7\sigma_{18} \text{ for an odd integer } x \text{ [85, Lemma 12.12]}. \tag{1.65}$$

In the sequel, we need:

$$\begin{array}{l}(1)\ H(\nu_4) = \iota_7,\ \text{[85, Lemma 5.4]}, \\ (2)\ H(\sigma') = \eta_{13} \text{ [85, Lemma 5.14]}, \\ (3)\ H(\sigma''') = 4\nu_9 \text{ [85, Lemma 5.1]}, \\ (4)\ H(\sigma_8) = \iota_{15} \text{ [85, Lemma 5.14]}, \\ (5)\ H(\varepsilon_3) = \nu_5^2 \text{ [85, Proposition 6.1]}, \\ (6)\ H(\mu_3) = \sigma''' \text{ [85, Lemma 6.5]}, \\ (7)\ H(\theta) = \eta_{23} \text{ for } \theta \in \{\sigma_{12}, \nu_{19}, \eta_{22}\}_1 \text{ [85, Lemma 7.5]}.\end{array} \tag{1.66}$$

Next, recall that by [85, Chap. VI, (ii)], an element of $\{\nu_6, \eta_9, \nu_{10}\}_t \in \pi_{14}^6/\{[\iota_6, \nu_6]\}$ for $0 \leq t \leq 4$ is denoted by $\bar{\nu}_6$. Then,

$$2\bar{\nu}_6 = \pm[\iota_6, \nu_6] \text{ [85, Lemma 6.2]}. \tag{1.67}$$

In the sequel, the following is useful as well:

$$\begin{aligned}&(1)\ \eta_5\bar{\nu}_6 = \nu_5^3\ [85, (7.3)],\\&(2)\ \eta_7\sigma_8 = \sigma'\eta_{14} + \bar{\nu}_7 + \varepsilon_7\ [85, (7.4)],\\&(3)\ \eta_6\sigma' = \sigma''\eta_{13} = 4\bar{\nu}_6\ [85, (7.4)]\end{aligned} \tag{1.68}$$

which, in view of [85, Lemma 6.4], lead to:

$$\begin{aligned}&(1)\ \eta_9\sigma_{10} = \bar{\nu}_9 + \varepsilon_9,\\&(2)\ \eta_n\sigma_{n+1} = \sigma_n\eta_{n+7} = \bar{\nu}_n + \varepsilon_n \text{ for } n \geq 10,\\&(3)\ \eta_6^2\sigma_8 = \nu_6^3 + \eta_6\varepsilon_7\end{aligned} \tag{1.69}$$

and, in view of [85, (7.1)], to

$$\sigma_9\eta_{16} + \bar{\nu}_9 + \varepsilon_9 = \sigma_9\eta_{16} + \eta_9\sigma_{10} = [\iota_9, \iota_9]. \tag{1.70}$$

Further, we need:

$$\begin{aligned}&(1)\ \sigma_{10}\varepsilon_{17} = \sigma_{10}\bar{\nu}_{17} = [\iota_{10}, \nu_{10}^2]\ [85, (10.20)],\\&(2)\ \varepsilon_3\sigma_{11} = 0\ [85, \text{Lemma } 10.7].\end{aligned} \tag{1.71}$$

To show next results, we need

$$\bar{\nu}_n\eta_{n+8} = \eta_n\bar{\nu}_{n+1} = \nu_n^3 \text{ for } n \geq 6\ [85, \text{Lemma } 6.3]. \tag{1.72}$$

Next, we need

Lemma 1.18. *Let* $n \equiv 3 \pmod 4$ *and* $n \geq 7$. *Then,*

1. $\{\Delta\iota_n, \eta_{n-1}, 2\iota_n\} = 0$;
2. $\Delta(E\{\eta_{n-1}, 2\iota_n, \alpha\}) = 0$, *where* $\alpha \in \pi_k(\mathbb{S}^n)$ *is an element satisfying* $2\iota_n \circ \alpha = 0$.

Proof. By [85, Proposition 1.4] and the fact that $2\pi_{n+1}(SO(n+1)) = 0$ [37, p. 161], we obtain

$$i_{n+1}(\mathbb{R})\circ\{\Delta\iota_n, \eta_{n-1}, 2\iota_n\} = -\{i_{n+1}(\mathbb{R}), \Delta\iota_n, \eta_{n-1}\}\circ 2\iota_{n+1} \subseteq 2\pi_{n+1}(SO(n+1)) = 0.$$

It follows from $(\mathcal{SO}^n_{n+1})$ and (1.55) that $i_{n+1}(\mathbb{R})_*\colon \pi_{n+1}(SO(n)) \to \pi_{n+1}(SO(n+1))$ is a monomorphism. This leads to (1).

By (1.54) and (1), for any $\beta \in \{\eta_{n-1}, 2\iota_n, \alpha\}$, we obtain

$$\Delta(E\beta) \in \Delta\iota_n \circ \{\eta_{n-1}, 2\iota_n, \alpha\} = -\{\Delta\iota_n, \eta_{n-1}, 2\iota_n\} \circ E\alpha = 0.$$

This leads to (2) and completes the proof. □

In view of (1.52), (1.57)(2), (1.60), and Lemma 1.18(2), we get

Example 1.19. $\Delta\varepsilon_n = 0$ and $\Delta\mu_n = 0$, if $n \equiv 3 \pmod 4$.

Next, we show

Lemma 1.20. 1. $\Delta(\nu_n^2) = 0$, *if* $n \equiv 5 \pmod 8$;
2. $\Delta(\nu_{4n}^2) = 0$, *if* n *is odd.*

Proof. Since $\pi_7(SO(5)) \cong \mathbb{Z}$ [37, p. 162], $\Delta\colon \pi_8(\mathbb{S}^5) \to \pi_7(SO(5))$ is trivial and $\Delta\nu_5 = 0$. Then, by (1.54), $\Delta(\nu_5^2) = 0$. Let now $n \equiv 5 \pmod 8$ and $n \geq 13$. We consider the exact sequence $(\mathcal{SO}^n_{n+5})$:

$$\pi_{n+6}(\mathbb{S}^n) \xrightarrow{\Delta} \pi_{n+5}(SO(n)) \xrightarrow{i_*} \pi_{n+5}(SO(n+1)) \to 0.$$

By [12, Theorem 2], we obtain

$$\pi_{n+5}(SO(n)) \cong \pi_{n+5}(SO) \oplus \pi_{n+6}(V_{n+8,8}).$$

In light of [30, Table 1], $\pi_{n+6}(V_{n+8,8}) \cong \mathbb{Z}_8$ and by [14], $\pi_{n+5}(SO) = 0$. So, $\pi_{n+5}(SO(n)) \cong \mathbb{Z}_8$. By [37, p. 161], $\pi_{n+5}(SO(n+1)) \cong \mathbb{Z}_8$. From the fact that $\pi_{n+6}(\mathbb{S}^n) = \{\nu_n^2\} \cong \mathbb{Z}_2$, we obtain $\Delta(\nu_n^2) = 0$, and hence (1) follows.

We obtain $\pi_9(SO(4)) \cong \pi_9(SO(3)) \oplus \pi_9(\mathbb{S}^3) \cong (\mathbb{Z}_3)^2$, and so $\Delta(\nu_4^2) = 0$. Let now $n \geq 3$. Then, we consider the exact sequence $(\mathcal{SO}^{4n}_{4n+5})$:

$$\pi_{4n+6}(\mathbb{S}^{4n}) \xrightarrow{\Delta} \pi_{4n+5}(SO(4n)) \xrightarrow{i_*} \pi_{4n+5}(SO(4n+1)) \to 0.$$

By [37, p. 161],

$$\pi_{4n+5}(SO(4n+1)) \cong \mathbb{Z}_2 \ (n \geq 2). \tag{1.73}$$

By [36, Theorem 1(iii)], $\pi_{17}(SO(12)) = \{[\iota_7]_{12}\eta_7\mu_8\} \cong \mathbb{Z}_2$. Since $J([\iota_7]_{12}\eta_7\mu_8) = \sigma_{12}\eta_{19}\mu_{20} \neq 0$ in $\pi_{29}(\mathbb{S}^{12})$, we get that $\Delta(\nu_{12}^2) = 0$. Let n be odd and $n \geq 5$. In light of [12, Theorem 2],

$$\pi_{4n+5}(SO(4n)) \cong \pi_{4n+5}(SO) \oplus \pi_{4n+6}(V_{4n+8,8}).$$

By means of [14] and [30, Table 1], $\pi_{4n+5}(SO) \cong \mathbb{Z}_2$ and $\pi_{4n+6}(V_{4n+8,8}) = 0$. Hence, we obtain $\Delta(\nu_{4n}^2) = 0$ if n is odd with $n \geq 5$. This leads to (2) and completes the proof. □

Kristensen and Madsen [38, Theorem 1.3] suggests the non-triviality of $[\iota_n, \nu_n^2]$ for $n \equiv 0, 1, 2, 3, 6 \pmod 8$ and $n \geq 6$ and [68, Proposition 3.4] gives an explicit proof of its non-triviality for $n \equiv 2 \pmod 4$ and $n \geq 6$.

By Lemma 1.2(1) and (1.32), we have $[\iota_n, \nu_n^2] = 0$ if $n \equiv 7 \pmod 8$ or $n = 2^i - 3$ for $i \geq 3$. In virtue of Lemma 1.20 and (1.6), we get that

$$[\iota_n, \nu_n^2] = 0, \ \text{if}\, n \equiv 5 \pmod 8 \tag{1.74}$$

and

$$[\iota_n, \nu_n^2] = 0, \ \text{if}\, n \equiv 4 \pmod 8. \tag{1.75}$$

Let now $n \equiv 0 \pmod 4$ and $n \geq 8$. By [12, 14, Theorem 2] and [30, Table 1], $\pi_{2n+3}(SO(2n-2)) \cong \mathbb{Z} \oplus \mathbb{Z}_4$. In the exact sequence $(\mathcal{SO}^{2n-3}_{2n+3})$, the map $p_{2n-2}(\mathbb{R})_*\colon \pi_{2n+3}(SO(2n-2)) \to \pi_{2n+3}(\mathbb{S}^{2n-3})$ is an epimorphism by Lemma 1.20(1). So, the direct summand $\mathbb{Z}_4$ of $\pi_{2n+3}(SO(2n-2))$ is generated by $[\nu^2_{2n-3}]$. By [37, p. 161], $\pi_{2n+3}(SO(2n+1)) \cong \mathbb{Z}\oplus\mathbb{Z}_2$ and $\pi_{2n+3}(SO(2n+2)) \cong \mathbb{Z}$. It follows from $(\mathcal{SO}^{2n+1}_{2n+3})$ that the direct summand $\mathbb{Z}_2$ of $\pi_{2n+3}(SO(2n+1))$ is generated by $\Delta\nu_{2n+1}$. By [37, p. 161], $\pi_{2n+3}(SO(2n+k-1)) \cong \mathbb{Z} \oplus \mathbb{Z}_2$ for $0 \leq k \leq 2$. Hence, by the use of $(\mathcal{SO}^{2n+k-1}_{2n+3})$ for $-1 \leq k \leq 2$, $(i_{2n-2,2n+1})_*\colon \pi_{2n+3}(SO(2n-2)) \to \pi_{2n+3}(SO(2n+1))$ is an epimorphism and we get the relation

$$[\nu^2_{2n-3}]_{2n+1} = \Delta\nu_{2n+1}.$$

Thus, we conclude

Lemma 1.21. *$E^3J[\nu^2_{2n-3}] = [\iota_{2n+1}, \nu_{2n+1}]$, if $n \equiv 0 \pmod 4$ and $n \geq 8$.*

Hereafter, we use often the EHP-sequence of the following type:

$$(\mathcal{PE}^n_{n+k}) \quad \pi^{2n+1}_{n+k+2} \xrightarrow{P} \pi^n_{n+k} \xrightarrow{E} \pi^{n+1}_{n+k+1}.$$

By [85, Proposition 2.5] and (1.5), we obtain

$$HP(E^3\gamma) = \pm(1 + (-1)^n)E\gamma \text{ for } \gamma \in \pi^{2n-2}_k. \tag{1.76}$$

Suppose that $\Delta\alpha = 0$ for $\alpha \in \pi_k(\mathbb{S}^{n-1})$. Then, the following is essentially proved in [91]; the proof is based on a generalization of [90, Theorem 5-1], with the sign corrected as in [91]:

$$H(J[\alpha]) = -E^n\alpha. \tag{1.77}$$

Now, we show

I. $[\iota_n, \nu^2_n] \neq 0$ if $n \equiv 1 \pmod 8$ and $n \geq 9$.

In virtue of (1.21) and [85, Lemmas 9.2, 10.1, Theorem 20.3], $[\iota_9, \nu^2_9] = \bar{\nu}_9\nu^2_{17} \equiv 2\kappa_9 + 8a\sigma^2_9 \neq 0$ for $a \in \{0, 1\}$.

Let $n \equiv 0 \pmod 4$ and $n \geq 8$. By Lemma 1.21, $[\iota_{2n+1}, \nu^2_{2n+1}] = E^3(J[\nu^2_{2n-3}] \circ \nu_{4n+1})$. Suppose that $E^3(J[\nu^2_{2n-3}] \circ \nu_{4n+1}) = 0$. Then, by the use of $(\mathcal{PE}^{2n}_{4n+6})$, we obtain $E^2(J[\nu^2_{2n-3}] \circ \nu_{4n+1}) = 8a[\iota_{2n}, \sigma_{2n}]$ for $a \in \{0, 1\}$. By means of [85, Proposition 11.11(i)], there exists an element $\beta \in \pi^{2n-2}_{4n+4}$ such that $P(8\sigma_{4n+1}) = E^2\beta$ and $H\beta \in \{2\iota_{4n-5}, \eta_{4n-5}, 8\sigma_{4n-4}\}_2$. By [85, (1.15), Proposition 1.2.0(ii), Lemma 1.1] and the relation $2\eta_{4n-5} = 0$, we see that

$$\{2\iota_{4n-5}, \eta_{4n-5}, 8\sigma_{4n-4}\}_2 \subseteq \{2\iota_{4n-5}, \eta_{4n-5}, 8\sigma_{4n-4}\} \subseteq$$

$$\{2\iota_{4n-5}, 0, 4\sigma_{4n-4}\} = 2\iota_{4n-5} \circ \pi^{4n-5}_{4n+4} + \pi^{4n-5}_{4n-3} \circ 4\sigma_{4n-3} = 0.$$

So, there exists an element $\beta' \in \pi^{2n-3}_{4n+3}$ such that $\beta = E\beta'$. Hence, $E^2(J[\nu^2_{2n-3}] \circ \nu_{4n+1}) = aE^3\beta'$.

In virtue of Lemma 1.2(1) and (1.15), $[\iota_{2n-1}, \eta_{2n-1}\sigma_{2n}] = 0$. In light of (1.6) and Example 1.19, $[\iota_{2n-1}, \varepsilon_{2n-1}] = 0$, and so $P\pi^{4n-1}_{4n+7} = 0$. Therefore, by $(\mathcal{PE}^{2n-1}_{4n+5})$, $E(J[\nu^2_{2n-3}] \circ \nu_{4n+1}) = aE^2\beta'$. Finally, by the use of $(\mathcal{PE}^{2n-2}_{4n+4})$ and (1.77), we have a contradictory relation $\nu^3_{4n-5} = 0$. Thus, we get $[\iota_{2n+1}, \nu^2_{2n+1}] = E^3(J[\nu^2_{2n-3}] \circ \nu_{4n+1}) \neq 0$.

We denote by $\mathbb{R}P^n$ the real n-dimensional projective space, by $\gamma_n\colon \mathbb{S}^n \to \mathbb{R}P^n$ the covering map and by $p'_n\colon \mathbb{R}P^n \to \mathbb{S}^n$ the collapsing map, respectively. Then, we can take $\Delta\iota_n = j \circ \gamma_{n-1}$, where $j\colon \mathbb{R}P^{n-1} \hookrightarrow SO(n)$ is the canonical embedding. Hence, by the relations $j \circ p_n(\mathbb{R}) = p'_{n-1}$ and $p'_n \circ \gamma_n = (1 + (-1)^{n+1})\iota_n$, we obtain

$$p_n(\mathbb{R})(\Delta\iota_n) = (1 + (-1)^n)\iota_{n-1}. \tag{1.78}$$

Let $n \equiv 0 \pmod 8$ and $n \geq 8$. By the use of $(\mathcal{SO}^{n-1}_{n+1})$ and [37, pp. 161–162], we get that $i_n(\mathbb{R})_*\colon \pi_{n+1}(SO(n-1)) \to \pi_{n+1}(SO(n))$ is a monomorphism. So, we obtain

$$\Delta\nu_{n-1} = 0, \text{ if } n \equiv 0 \pmod 8 \text{ and } n \geq 8. \tag{1.79}$$

Hence, by Lemma 1.20(2), ν_{n-1} and ν^2_{n-4} are lifted to $[\nu_{n-1}] \in \pi_{n+2}(SO(n))$ and $[\nu^2_{n-4}] \in \pi_{n+2}(SO(n-3))$, respectively. We show the following

Lemma 1.22. *Let* $n \equiv 0 \pmod 8$ *and* $n \geq 16$. *Then,*

1. $2[\nu_{n-1}] - \Delta\nu_n = x[\nu^2_{n-4}]_n$ *for odd* x;
2. $\pi_{n+5}(SO(n+1)) = \{[\nu_{n-1}]_{n+1}\nu_{n+2}\} \cong \mathbb{Z}_2$.

Proof. By the use of $(\mathcal{SO}^{n-k}_{n+2})$ for $2 \leq k \leq 4$, Lemma 1.20 and [37, p. 161], we see that $(i_{n-3,n-1})_*\colon \pi_{n+2}(SO(n-3)) \to \pi_{n+2}(SO(n-1)) \cong \mathbb{Z}_8$ is an isomorphism and $\pi_{n+2}(SO(n-3)) = \{[\nu^2_{n-4}]\}$. In virtue of [37, p. 161], $\pi_{n+2}(SO(n+1)) \cong \mathbb{Z}_8$ and $\pi_{n+2}(SO(n)) \cong \mathbb{Z}_{24} \oplus \mathbb{Z}_8$. Then, by $(\mathcal{SO}^{n-k}_{n+2})$ for $k = 0, 1$, we get $\pi_{n+2}(SO(n)) = \{\Delta\nu_n, [\nu_{n-1}]\}$. By (1.78), we obtain $p_n(\mathbb{R})(\Delta\nu_n) = 2\nu_{n-1}$, and hence $2[\nu_{n-1}] - \Delta\nu_n \in \operatorname{Im}\{i_n(\mathbb{R})_*\colon \pi_{n+2}(SO(n-1)) \to \pi_{n+2}(SO(n))\}$. Since $\sharp(2[\nu_{n-1}] - \Delta\nu_n) = 8$, we have the required relation of (1).

We consider the exact sequence $(\mathcal{SO}^n_{n+5})$:

$$\pi_{n+6}(S^n) \xrightarrow{\Delta} \pi_{n+5}(SO(n)) \xrightarrow{i_*} \pi_{n+5}(SO(n+1)) \longrightarrow 0.$$

By (1.73), $\pi_{n+5}(SO(n+1)) \cong \mathbb{Z}_2$. In view of [12, 14, Theorem 2] and [30, Table 1], we obtain

$$\pi_{n+5}(SO(n)) \cong (\mathbb{Z}_2)^2 \ (n \equiv 0 \pmod 8 \text{ and } n \geq 8). \tag{1.80}$$

By (1.79), ν^2_{n-1} is lifted to $[\nu_{n-1}]\nu_{n+2}$. Consequently, we obtain $\pi_{n+5}(SO(n)) = \{\Delta(\nu^2_n), [\nu_{n-1}]\nu_{n+2}\}$ and $\pi_{n+5}(SO(n+1)) = \{[\nu_{n-1}]_{n+1}\nu_{n+2}\}$. This leads to (2) and completes the proof. □

The relation in [85, Lemma 11.17] is regarded as the J-image of that in Lemma 1.22(1).

Further,

$$\eta_5^2\varepsilon_7 = 4\nu_5\sigma_8 \neq 0 \text{ [85, Lemma 6.6, Theorem 7.3 and (7.10)]}, \tag{1.81}$$

$$\begin{array}{l}(1)\ \sigma'\nu_{14} = x\nu_7\sigma_{10}, \\ (2) \pm P(\nu_{17}) = 2\sigma_8\nu_{15} - x\nu_8\sigma_{11}\end{array} \text{ for an odd integer } x \text{ [85, (7.19)]} \tag{1.82}$$

and

$$\nu_{10}\sigma_{13} = 2\sigma_{10}\nu_{17} = [\iota_{10}, \eta_{10}] \text{ [85, (7.19) and (7.21)]} \tag{1.83}$$

which leads to

$$\nu_{11}\sigma_{14} = 0. \tag{1.84}$$

We also recall the result

$$\sigma_{11}\nu_{18} = [\iota_{11}, \iota_{11}] \text{ [85, (7.21)]} \tag{1.85}$$

which yields

$$\sigma_n\nu_{n+7} = 0 \text{ for } n \geq 12. \tag{1.86}$$

Remark 1.23. The results in (1.53), (1.55), Lemma 1.20, Example 1.19, and (1.79) overlap with [30, Table 2].

Now, we present a proof of the non-triviality of $[\iota_n, \nu_n^2]$ in the case $n \equiv 0 \pmod 8$ and $n \geq 8$.

II. $[\iota_n, \nu_n^2] \neq 0$ if $n \equiv 0 \pmod 8$ and $n \geq 8$.

By (1.22) and [85, Theorem 7.7], $[\iota_8, \nu_8^2] = \nu_8\sigma_{11}\nu_{18} \neq 0$. Let $n \equiv 0 \pmod 8$ and $n \geq 16$. In light of (1.80), $\pi_{n+5}(SO(n)) \cong (\mathbb{Z}_2)^2$. Then, by (1.54) and Lemma 1.22,

$$\Delta(\nu_n^2) = [\nu_{n-4}^2]_n\nu_{n+2}$$

and hence $[\iota_n, \nu_n^2] = E^3(J[\nu_{n-4}^2] \circ \nu_{2n-1})$.

This result is written in Nomura [71] by using a different notation. Suppose that $E^3(J[\nu_{n-4}^2] \circ \nu_{2n-1}) = 0$. Then, $E^2(J[\nu_{n-4}^2] \circ \nu_{2n-1}) \in P\pi_{2n+6}^{2n-1} = \{[\iota_{n-1}, \sigma_{n-1}]\}$. By [85, Proposition 11.11(ii)], it holds $P\pi_{2n+5}^{2n-3} \subseteq E^2\pi_{2n+1}^{n-4}$. Then, by (1.43) and using $(\mathcal{PE}_{2n+4-k}^{n-1-k})$ for $k = 0, 1$, we get that

$$J[\nu_{n-4}^2] \circ \nu_{2n-1} - aE^5(\gamma\sigma_{2n-10}) - E\beta \in P\pi_{2n+4}^{2n-5}$$

for some $\beta \in \pi_{2n+1}^{n-4}$ and $a \in \{0, 1\}$. Hence, (1.76) and (1.77) imply a contradictory relation $\nu_{2n-7}^3 = 0$, and thus $[\iota_n, \nu_n^2] \neq 0$.

The result **II** is essentially obtained by Nomura [71].

1.4 Proof of Theorem 1.14, Part II

Let $\tau'_{2n} = r_n\omega_{n,\mathbb{C}} \in \pi_{2n}(SO(2n))$ and $\bar{\tau}'_{4n} = r_{2n}c_n\omega_{n,\mathbb{H}} \in \pi_{4n+2}(SO(4n))$.
It is well known that

$$p_{2n}(\mathbb{R})\tau'_{2n} = (n-1)\eta_{2n-1} \text{ and } p_{4n}(\mathbb{R})\bar{\tau}'_{4n} = \pm(n+1)\nu_{4n-1} \text{ for } n \geq 2. \quad (1.87)$$

Whence, by the commutative diagram

$$\begin{array}{ccc}
\pi_{4n+2}(U(2n)) & \xrightarrow{i_{2n+1}(\mathbb{C})_*} & \pi_{4n+2}(U(2n+1)) \\
\downarrow{\scriptstyle r_{2n*}} & & \downarrow{\scriptstyle r_{2n+1*}} \\
\pi_{4n+2}(SO(4n)) & \xrightarrow{i_{4n,4n+2*}} & \pi_{4n+2}(SO(4n+2)),
\end{array}$$

we obtain

Lemma 1.24. 1. *If n is even and $n \geq 4$ then $i_{n+1}(\mathbb{R})\tau'_n = \Delta\iota_{n+1}$ and $p_n(\mathbb{R})\tau'_n = (\frac{n}{2}-1)\eta_{n-1}$;*
2. *If $n \equiv 0 \pmod 4$ and $n \geq 8$ then $(i_{n,n+2})\bar{\tau}'_n = \tau'_{n+2}$ and $p_n(\mathbb{R})\bar{\tau}'_n = \pm(\frac{n}{4}+1)\nu_{n-1}$.*

Let $n \equiv 2 \pmod 4$ and $n \geq 10$. Then, by the use of $(\mathcal{SO}^n_n)$, Lemma 1.24(1) and [37, p. 161], we obtain

$$\pi_n(SO(n)) = \{\tau'_n\} \cong \mathbb{Z}_4 \text{ and } 2\tau'_n = \Delta\eta_n, \text{ if } n \equiv 2 \pmod 4 \text{ and } n \geq 10. \quad (1.88)$$

By the use of $(\mathcal{SO}^{4n+1}_{4n+2})$, Lemma 1.24(1) and [37], we obtain

$$\Delta(\eta^2_{4n+1}) = 4i_{4n+1}(\mathbb{R})\bar{\tau}'_{4n}, \text{ if } n \geq 2. \quad (1.89)$$

Then, by Lemma 1.24(1), $(\mathcal{SO}^{4n}_{4n+2})$ and (1.89), we have $\tau'_{4n}\eta^2_{4n} - 4\bar{\tau}'_{4n} \in \{\Delta\nu_{4n}\}$. Composing $p_{4n}(\mathbb{R})$ with this relation, using the fact that $\eta^3_{4n-1} = 12\nu_{4n-1}$ (1.18), (1.78) and (1.87),

$$\tau'_{4n}\eta^2_{4n} \equiv 4\bar{\tau}'_{4n} \pmod{2a\Delta\nu_{4n}}, \text{ for } a \text{ odd and } n \geq 2.$$

Set $\tau_{2n} = J\tau'_{2n} \in \pi_{4n}(\mathbb{S}^{2n})$ and $\bar{\tau}_{4n} = J\bar{\tau}'_{4n} \in \pi_{8n+2}(\mathbb{S}^{4n})$. Then, we note that

$$E\tau_{2n} = [\iota_{2n+1}, \iota_{2n+1}], H\tau_{2n} = (n-1)\eta_{4n-1} \quad (1.90)$$

and

$$E^3\bar{\tau}_{4n} = [\iota_{4n+3}, \iota_{4n+3}], H\bar{\tau}_{4n} = \pm(n+1)\nu_{8n-1} \quad (1.91)$$

By (1.89), we have

$$4E\bar{\tau}_{4n} = [\iota_{4n+1}, \eta^2_{4n+1}]. \tag{1.92}$$

We notice that (1.91) overlaps with [70, Lemma 6.3] and (1.92) overlaps with [70, Lemma 5.3].

In the sequel, we need:

Lemma 1.25. *If the map* $\eta^*_{n-1}i^*_n : [M^n, X] \to \pi_{n-1}(X)$ *is an epimorphism then*

$$\pi_n(X) \circ p_n = 2[M^n, X]$$

for $n \geq 3$. *In particular, this holds provided* $\eta^*_{n-1} : \pi_{n-1}(X) \to \pi_n(X)$ *is an epimorphism and* $2\pi_{n-1}(X) = 0$ *or* $\eta^*_{n-1} : \pi_{n-1}(X) \to \pi_n(X)$ *is an isomorphism for* $n \geq 4$.

Proof. If $\eta^*_{n-1}i^*_n : [M^n, X] \to \pi_{n-1}(X)$ is an epimorphism then $\pi_n(X) \circ p_n = [M^n, X] \circ (i_n\eta_{n-1}p_n)$ and the relation (1.40) yields that $\pi_n(X) \circ p_n = 2[M^n, X]$.

Next, $\alpha \in \pi_{n-1}(X)$ satisfying $2\alpha = 0$ is extendible to $\bar{\alpha} \in [M^n, X]$. This implies that the map $i^*_n : [M^n, X] \to \pi_{n-1}(X)$ is an epimorphism.

Certainly, the isomorphism $\eta^*_{n-1} : \pi_{n-1}(X) \to \pi_n(X)$ and $2\eta_{n-1} = 0$ lead to $2\pi_{n-1}(X) = 0$. Consequently, in view of the first part, the proof is complete. □

Let $\bar{\eta}_n \in [M^{n+2}, \mathbb{S}^n]$ and $\tilde{\eta}_n \in \pi_{n+2}(M^{n+1})$ be an extension and a coextension of η_n with $\bar{\eta}_n i_{n+2} = \eta_{n+1}$ and $p_{n+1}\tilde{\eta}_n = \eta_{n+1}$ for $n \geq 3$, respectively. We note that

$$\bar{\eta}_n \in \{\eta_n, 2\iota_{n+1}, p_{n+1}\}, \text{ if } n \geq 3 \tag{1.93}$$

and

$$\tilde{\eta}_n \in \{i_{n+1}, 2\iota_n, \eta_n\}, \text{ if } n \geq 3. \tag{1.94}$$

Further, by [63, Lemma 4.1], there is a lift $\tilde{\eta}_2 \in \pi_4(M^3)$ of η_3 satisfying $E\tilde{\eta}_2 = \tilde{\eta}_3$, $2\tilde{\eta}_2 = i_3\eta^2_2$ and $p_3\tilde{\eta}_2 = \eta_3$. Then, in view of (1.40), we have

$$2\bar{\eta}_n = \eta^2_n p_{n+2} \text{ for } n \geq 3 \text{ and } 2\tilde{\eta}_n = i_{n+1}\eta^2_n \text{ for } n \geq 2. \tag{1.95}$$

Notice that (1.93) and (1.94) lead to an alternative proof of (1.95).

Namely, $2\bar{\eta}_n \in \{\eta_n, 2\iota_{n+1}, p_{n+1}\} \circ 2\iota_{M^{n+1}} = \{\eta_n, 2\iota_{n+1}, p_{n+1}\} \circ i_{n+1} \circ \eta_n \circ p_{n+1} = \eta_n \circ \{2\iota_{n+1}, p_{n+1}, i_{n+1}\} \circ \eta_{n+1} \circ p_{n+2} = \eta^2_n p_{n+2}$, because $\{2\iota_{n+1}, p_{n+1}, i_{n+1}\} \ni \iota_n \pmod{2\iota_n}$ and, in view of (1.17)(1), it holds $2\tilde{\eta}_n \in \{i_{n+1}, 2\iota_n, \eta_n\} \circ 2\iota_{n+2} = i_{n+1} \circ \{2\iota_n, \eta_n, 2\iota_{n+1}\} = i_{n+1}\eta^2_n$.

Next, the relation (1.17)(2) yields

$$\bar{\eta}_n\tilde{\eta}_{n+1} = \pm E^{n-3}\nu' \text{ for } n \geq 3 \tag{1.96}$$

and (1.95) implies:

Lemma 1.26. $\pi_{n+1}(M^n) = \mathbb{Z}_4\{\tilde{\eta}_{n-1}\}$ *and* $[M^{n+2}, \mathbb{S}^n] = \mathbb{Z}_4\{\tilde{\eta}_n\}$ *for* $n \geq 3$.

To state a next result, we recall that $V_{n,k} = SO(n)/SO(n-k)$ denotes the Stiefel manifold and, by [63, Proposition 2.1], there is the cell structure $V_{2n+1,2} = M^{2n} \cup_{\lambda_n} e^{4n-1}$ for some $\lambda_n \in \pi_{4n-2}(M^{2n})$ for $n \geq 2$, where the order of λ_n is 4 for n even and 8 for n odd [62, Lemma 12]. Further,

$$E\lambda_n = i_{2n+1} \circ [\iota_{2n}, \iota_{2n}] \neq 0 \text{ [62, (6)] and [67, Lemma 1.2].} \tag{1.97}$$

Lemma 1.27. 1. $\pi_6(M^4) = \mathbb{Z}_4\{\lambda_2\} \oplus \mathbb{Z}_2\{\tilde{\eta}_3\eta_5\}$, $\pi_7(M^5) = \mathbb{Z}_2\{\tilde{\eta}_4\eta_6\} \oplus \mathbb{Z}_4\{i_5\nu_4\}$ *and* $\pi_{n+2}(M^n) = \mathbb{Z}_2\{\tilde{\eta}_{n-1}\eta_{n+1}\} \oplus \mathbb{Z}_2\{i_n\nu_{n-1}\}$ *for* $n \geq 6$;
2. $\pi_7(M^4) = \mathbb{Z}_2\{\lambda_2\eta_6\} \oplus \mathbb{Z}_2\{\tilde{\eta}_3\eta_5^2\}$, $\pi_{n+3}(M^n) = \mathbb{Z}_2\{\tilde{\eta}_{n-1}\eta_{n+1}^2\} \oplus \mathbb{Z}_2\{i_n\nu_{n-1}\eta_{n+2}\}$ *for* $n = 5, 6$ *and* $\pi_{n+3}(M^n) = \mathbb{Z}_2\{\tilde{\eta}_{n-1}\eta_{n+1}^2\}$ *for* $n \geq 7$;
3. $\pi_9(M^5) = \mathbb{Z}_2\{i_5\nu_4\eta_5^2\} \oplus \mathbb{Z}_2\{[\tilde{\eta}_4, i_5]\}$, $\pi_{10}(M^6) = \mathbb{Z}_8\{\lambda_3\}$, $\pi_{11}(M^7) = \mathbb{Z}_2\{E\lambda_3\}$ *and* $\pi_{n+4}(M^n) = 0$ *for* $n \geq 8$.

Proof. 1. In view of [63, Lemma 2.2] it holds $\pi_6(M^4) = \mathbb{Z}_4\{\lambda_2\} \oplus \mathbb{Z}_2\{\tilde{\eta}_3\eta_5\}$ and [61, Lemma 3.4] yields $\pi_7(M^5) = \mathbb{Z}_2\{\tilde{\eta}_4\eta_6\} \oplus \mathbb{Z}_4\{i_5\nu_4\}$. Next, by [59, 63] it holds $\pi_{n+2}(M^n) = \mathbb{Z}_2\{\tilde{\eta}_{n-1}\eta_{n+1}\} \oplus \mathbb{Z}_2\{i_n\nu_{n-1}\}$ for $n \geq 6$.
2. By virtue of [63, Lemma 2.2], we get $\pi_7(M^4) = \mathbb{Z}_2\{\lambda_2\eta_6\} \oplus \mathbb{Z}_2\{\tilde{\eta}_3\eta_5^2\}$, $\pi_8(M^5) = \mathbb{Z}_2\{i_5\nu_4\eta_7\} \oplus \mathbb{Z}_2\{\tilde{\eta}_4\eta_6^2\}$ and $\pi_9(M^6) = \mathbb{Z}_2\{\tilde{\eta}_5\eta_7^2\} \oplus \mathbb{Z}_2\{i_6\nu_5\eta_8\}$. Further, [59] yields $\pi_{n+3}(M^n) = \mathbb{Z}_2\{\tilde{\eta}_{n-1}\eta_{n+1}^2\}$ for $n \geq 7$.
3. In view of [63, Lemma 2.2], it holds $\pi_9(M^5) = \mathbb{Z}_2\{i_5\nu_4\eta_5^2\} \oplus \mathbb{Z}_2\{[\tilde{\eta}_4, i_5]\}$. Next, by means of [76, p. 266], we have $\pi_{10}(V_{7,2}) = 0$. Hence, [63, Proposition 2.1(i)] yields that $\pi_{10}(M^6) = \mathbb{Z}_8\{\lambda_3\}$. But [58, Lemma 2.7] or [62, Lemma 12] and (1.97) lead to $\pi_{11}(M^7) = \mathbb{Z}_2\{E\lambda_3\}$. Next, the homotopy exact sequence of the pair $(M^n, \mathbb{S}^{n-1})$ leads to $\pi_{n+4}(M^n) = 0$ for $n \geq 8$ and the proof is complete.

□

Next, we recall from [86] that $M^n \wedge M^n$ is a mapping cone of $2\iota_{M^{2n-1}} = i_{2n-1}\eta_{2n-2}p_{2n-1}$. Let $i'_4 : M^4 \hookrightarrow E(M^2 \wedge M^2)$ be the inclusion map. We set $i'_n = E^{n-4}i'_4 : M^n \hookrightarrow E^{n-3}(M^2 \wedge M^2)$ for $n \geq 4$ and $i''_n = i'_n i_n$. The Toda bracket $\{i'_4, 2i'_4, i_4\} \subseteq \pi_4(E(M^2 \wedge M^2))$ is well defined and its representative $\tilde{i}_4$ is a coextension of i_4 with $2\tilde{i}_4 = i''_4\eta_3$. If we set $\tilde{i}_n = E^{n-4}\tilde{i}_4 \in \pi_n(E^{n-3}(M^2 \wedge M^2))$ then $2\tilde{\iota}_{2n} = E(i_n \wedge i_n)\eta_{2n-1} = i''_{2n}\eta_{2n-1}$ for $n \geq 3$. Thus,

$$H(\lambda_n) = \pm\tilde{i}_{4n-2} \text{ [62, Proposition 14]} \tag{1.98}$$

for $H : \pi_{4n-2}(M^{2n}) \to \pi_{4n-2}(E(M^{2n-1} \wedge M^{2n-1}))$. Further,

$$P(\tilde{\iota}_{4n}) = \pm 2\lambda_n \text{ [62, Lemma 12]} \tag{1.99}$$

for $P : \pi_{4n}(E(M^{2n} \wedge M^{2n})) \to \pi_{4n-2}(M^{2n})$.
Then, (1.99) yields

$$P(\tilde{\iota}_{4n}\eta_{4n}) = P(\tilde{\iota}_{4n})\eta_{4n-2} = \pm 2\lambda_n\eta_{4n-2} = 0.$$

Recall also that in view of [58, Lemma 1.5(i)] and the methods of its proof, we have:

$$\begin{aligned}(1)\ &\pi_{2n-1}(E(M^n \wedge M^n)) = \mathbb{Z}_2\{i''_{2n}\},\\ (2)\ &\pi_{2n}(E(M^n \wedge M^n)) = \mathbb{Z}_4\{\tilde{\imath}_{2n}\}\end{aligned} \tag{1.100}$$

for $n \geq 3$.

Further, in view of [64, Lemma 1.4], we have

$$\begin{aligned}(1)\ &P(i'_{2n}) = [\iota_{M^n}, i_n] \in [M^{2n-2}, M^n],\\ (2)\ &P(i''_{4n-2}) = [i_{2n-1}, i_{2n-1}]\end{aligned} \tag{1.101}$$

for $n \geq 4$.

Lemma 1.28. *The suspension map* $E : \pi_{4n-4}(M^{2n-1}) \to \pi_{4n-3}(M^{2n})$ *is an isomorphism for* $n \geq 2$.

Proof. Because (1.97) and (1.101)(2) lead to $P(i''_{4n-2}) = E(\lambda_{n-1}) \neq 0$, we deduce from (1.100)(1) that $P : \pi_{4n-3}(E(M^{2n-1} \wedge M^{2n-1})) \to \pi_{4n-5}(M^{2n-1})$ is a monomorphism. Next, (1.98) and (1.100)(2) show that the map $P : \pi_{4n-2}(E(M^{2n-1} \wedge M^{2n-1})) \to \pi_{4n-4}(M^{2n-1})$ is trivial.

Now, the EHP-sequence

$$0 \to \pi_{4n-4}(M^{2n-1}) \xrightarrow{E} \pi_{4n-3}(M^{2n}) \xrightarrow{H} \pi_{4n-3}(E(M^{2n-1} \wedge M^{2n-1})) \xrightarrow{P} \pi_{4n-5}(M^{2n-1})$$

completes the proof. □

Remark 1.29. Notice that Lemma 1.28 overlaps with [15, Proposition 2.5(ii)] and [63, Lemma 4.1] for $n = 2$, and [63, Lemma 2.2(iv)] for $n = 3$.

The following result is useful:

Proposition 1.30 ([42, Proposition 5.2]). *For* $n \geq 2$, *there exists a generator* $\theta_{4n-1} \in \pi_{4n-1}(V_{2n+1,2})$ *such that* $p_*\theta_{4n-1} = a_n[\iota_{2n}, \iota_{2n}]$, *where* $p : V_{2n+1,2} \to \mathbb{S}^{4n-1}$ *is the projection map and* $a_n = 2$ *if* $n \equiv 1 \pmod 2$, *and* $a_n = 1$ *if* $n \equiv 0 \pmod 2$.

Now, by the fact that $\pi_{10}(M^6) = \mathbb{Z}_8\{\lambda_3\}$ and the cellular structure of $V_{7,2} = M^6 \cup_{\lambda_3} e^{11}$, we obtain $\pi_{10}(V_{7,2}) = 0$ [76, p. 266]. Hence, the fibration $\mathbb{S}^5 \xrightarrow{i} V_{7,2} \xrightarrow{p} \mathbb{S}^6$ leads to the exact sequence

$$\cdots \to \pi_{12}(\mathbb{S}^6) \xrightarrow{\Delta} \pi_{11}(\mathbb{S}^5) \xrightarrow{i_*} \pi_{11}(V_{7,2}) \xrightarrow{p_*} \pi_{11}(\mathbb{S}^6) \longrightarrow \pi_{10}(\mathbb{S}^5) \to 0.$$

Then, $\pi_{10}(\mathbb{S}^5) = \mathbb{Z}_2\{\nu_5\eta_8^2\}$ and $\pi_{11}(\mathbb{S}^6) = \mathbb{Z}\{[\iota_6, \iota_6]\}$ yield that $2[\iota_6, \iota_6] \in p_*\pi_{11}(V_{7,2})$. Because $\Delta(\nu_6^2) = 2\iota_5 \circ \nu_5^2 = 0$, we get that

$$\pi_{11}(V_{7,2}) = \mathbb{Z}_2\{i_*\nu_5^2\} \oplus \mathbb{Z}\{[2[\iota_6, \iota_6]]\},$$

where $[2[\iota_6, \iota_6]]$ is a lift of $2[\iota_6, \iota_6]$.

We notice that the result $[2[\iota_6,\iota_6]] \in \pi_{11}(V_{7,2})$ is a direct consequence of Proposition 1.30 and $\pi_{11}(V_{7,2}) = \mathbb{Z}_2\{i_*\nu_5^2\} \oplus \mathbb{Z}\{[2[\iota_6,\iota_6]]\}$ corresponds to the comments presented on [83, p. 132^{18-20}].

Next, consider the homotopy sequence

$$\cdots \to \pi_{12}(V_{7,2}) \xrightarrow{j'_*} \pi_{12}(V_{7,2}, M^6) \xrightarrow{\partial} \pi_{11}(M^6) \xrightarrow{i'_*} \pi_{11}(V_{7,2}) \xrightarrow{j'_*} \pi_{11}$$
$$(V_{7,2}, M^6) \xrightarrow{\partial} \pi_{10}(M^6) \to 0$$

of the pair $(V_{7,2}, M^6)$, where $i' : M^6 \to V_{7,2} = M^6 \cup_{\lambda_3} e^{11}$ and $j' : (V_{7,2}, *) \to (V_{7,2}, M^6)$ are the inclusion maps and write $\widetilde{\nu_5^2} \in \pi_{12}(M^6)$ for a coextension of the element $\nu_5^2 \in \pi_{11}(\mathbb{S}^5)$.

Although Jin-ho Lee has obtained that $\pi_{12}(V_{7,2}) \cong \mathbb{Z}_2$, we show:

Lemma 1.31. $\pi_{12}(V_{7,2}) = \mathbb{Z}_2\{i'\widetilde{\nu_5^2}\}$.

Proof. The exact sequence

$$\cdots \longrightarrow \pi_{12}(\mathbb{S}^5) \xrightarrow{i_*} \pi_{12}(V_{7,2}) \xrightarrow{p_*} \pi_{12}(\mathbb{S}^6) \to 0$$

of the fibration $\mathbb{S}^5 \xrightarrow{i} V_{7,2} \xrightarrow{p} \mathbb{S}^6$ leads to the short one

$$0 \to i_*\pi_{12}(\mathbb{S}^5) \longrightarrow \pi_{12}(V_{7,2}) \xrightarrow{p_*} \pi_{12}(\mathbb{S}^6) \to 0.$$

Because $p_*(i'\tilde{\nu}_5^2) = \nu_6^2$, the order $\sharp\tilde{\nu}_5^2 = 2$ [65, Lemma 5.2] and $\pi_{12}(\mathbb{S}^6) = \mathbb{Z}_2\{\nu_6^2\}$, we have

$$\pi_{12}(V_{7,2}) = i_*\pi_{12}(\mathbb{S}^5) \oplus \mathbb{Z}_2\{i'\widetilde{\nu_5^2}\}.$$

On the other hand, the exact sequence

$$\cdots \longrightarrow \pi_{12}(SO(7)) \longrightarrow \pi_{12}(V_{7,2}) \longrightarrow \pi_{11}(SO(5)) \longrightarrow \pi_{11}(SO(7)) \longrightarrow \cdots,$$

of the fibration $SO(5) \to SO(7) \to V_{7,2}$, the homotopy groups $\pi_{11}(SO(5)) \cong \mathbb{Z}_2$ and $\pi_{12}(SO(7)) = 0$ [46, Appendix A, Table VII, Topology, p. 1745] imply that $\pi_{12}(V_{7,2}) \cong \mathbb{Z}_2$ or 0.

Then, in view of the above, we derive that $\pi_{12}(V_{7,2}) = \mathbb{Z}_2\{i'\widetilde{\nu_5^2}\}$ and the proof is complete. □

Now, we are in a position to state:

Proposition 1.32. $\pi_{10}(M^5) = \mathbb{Z}_4\{i_5\nu_4^2\} \oplus \mathbb{Z}_2\{[\tilde{\eta}_4, i_5]\eta_9\}$, $\pi_{11}(M^6) = \mathbb{Z}_2\{\lambda_3\eta_{10}\} \oplus \mathbb{Z}_2\{i_6\nu_5^2\}$ *and* $\pi_{n+5}(M^n) = \mathbb{Z}_2\{i_n\nu_{n-1}^2\}$ *for* $n \geq 7$.

Proof. By means of [65, Theorem 1.2], we have $\pi_{10}(M^5) = \mathbb{Z}_4\{i_5\nu_4^2\} \oplus \mathbb{Z}_2\{[\tilde{\eta}_4, i_5]\eta_9\}$.

Since $j'_*i'_* = 0$, by means of Lemma 1.31, we get the triviality of $j'_* : \pi_{12}(V_{7,2}) \to \pi_{12}(V_{7,2}, M^6)$. Hence, the homotopy sequence of the pair $(V_{7,2}, M^6)$ becomes

$$0 \to \pi_{12}(V_{7,2}, M^6) \xrightarrow{\partial} \pi_{11}(M^6) \xrightarrow{i'_*} \pi_{11}(V_{7,2}) \xrightarrow{j'_*} \pi_{11}(V_{7,2}, M^6) \xrightarrow{\partial} \pi_{10}(M^6) \to 0.$$

In view of Blakers–Massey theorem (see, e.g., [92, Chap. VII, (7.12) Theorem]), $\pi_{12}(V_{7,2}, M^6) \cong \pi_{12}(\mathbb{S}^{11})$ and $\pi_{11}(V_{7,2}, M^6) \cong \pi_{11}(\mathbb{S}^{11})$. Then, the group structures of $\pi_{11}(V_{7,2})$ and $\pi_{10}(M^6)$ imply that $j'_*([2[\iota_6, \iota_6]]) = 8\hat{\iota}_{10}$, where $\hat{\iota}_{10}$ is the generator of $\pi_{11}(V_{7,2}, M^6)$ corresponding to ι_{11}. Further, $\pi_{12}(V_{7,2}, M^6) \xrightarrow{\partial} \pi_{11}(M^6)$ is a monomorphism and $\partial(\pi_{12}(V_{7,2}, M^6)) = \mathbb{Z}_2\{\lambda_3\eta_{10}\}$. Because $i_*(i_6\nu_5^2) = i'\nu_5^2$, the above yields $\pi_{11}(M^6) = \mathbb{Z}_2\{\lambda_3\eta_{10}\} \oplus \mathbb{Z}_2\{i_6\nu_5^2\}$.

Next, granting Lemma 1.28, the suspension map $E : \pi_{12}(M^7) \to \pi_{13}(M^8)$ is an isomorphism and, by means of the homotopy exact sequence of the pair $(M^n, \mathbb{S}^{n-1})$, we obtain $\pi_{n+5}(M^n) = \{i_n\nu_{n-1}^2\}$ for $n \geq 8$. This competes the proof. □

We point out that those homotopy groups $\pi_{n+k}(M^n)$ have also been computed in [93, Chap. 5] by means of other methods.

Now, we first recall:

$$\begin{aligned}
&(1)\ \{\nu_6, \eta_9, 2\iota_{10}\} = [\iota_6, \iota_6] + \{2[\iota_6, \iota_6]\} \text{ [85, Lemma 5.10]},\\
&(2)\ \{\nu_n, \eta_{n+3}, 2\iota_{n+4}\} = 0 \text{ for } n \geq 7,\\
&(3)\ \{2\iota_6, \eta_6, \nu_7\} = 2\pi_{11}(\mathbb{S}^6),\\
&(4)\ \{2\iota_7, \eta_7, \nu_8\} = 0,\\
&(5)\ \{2\iota_5, \eta_5, \nu_6\} = \nu_5\eta_8^2 \text{ [64, Remark (ii)]}
\end{aligned} \tag{1.102}$$

and then show:

Lemma 1.33. *$\varepsilon_n = \{\eta_n\bar{\eta}_{n+1}, \tilde{\eta}_{n+2}, \nu_{n+4}\}_{n-5}$ for $n \geq 5$.*

Proof. By (1.102)(4), the relation (1.94) for $n = 7$ and [85, Propositon 1.4], we get

$$\tilde{\eta}_7 \circ \nu_9 \in \{i_8, 2\iota_7, \eta_7\} \circ \nu_9 = i_8 \circ \{2\iota_7, \eta_7, \nu_8\} = 0.$$

Then, in view of (1.57)(1) and (1.96), we can take

$$\varepsilon_5 \in \{\eta_5, 2\nu_6, \nu_9\} = \{\eta_5, \bar{\eta}_6\tilde{\eta}_7, \nu_9\} = \{\eta_5\bar{\eta}_6, \tilde{\eta}_7, \nu_9\}$$

which lead to

$$\varepsilon_n = E^{n-5}\varepsilon_5 \in E^{n-5}\{\eta_5\bar{\eta}_6, \tilde{\eta}_7, \nu_9\} \subseteq \{\eta_n\bar{\eta}_{n+1}, \tilde{\eta}_{n+2}, \nu_{n+4}\}_{n-5} \text{ if } n \geq 5.$$

The indeterminacy of the Toda bracket $\{\eta_n\bar{\eta}_{n+1}, \tilde{\eta}_{n+2}, \nu_{n+4}\}$ is $\eta_n\bar{\eta}_{n+1} \circ \pi_{n+8}(M^{n+3}) + \pi_{n+5}(\mathbb{S}^n) \circ \nu_{n+5}$. Since $\eta_{n+4}\nu_{n+5} = 0$ (1.33) and $\pi_{n+5}(\mathbb{S}^n) = \{\nu_n\eta_{n+3}^2\}$ if $n \geq 5$, we obtain $\pi_{n+5}(\mathbb{S}^n) \circ \nu_{n+5} = 0$. Further, in view of Proposition 1.32, we have $\pi_{n+8}(M^{n+3}) = \{i_{n+3}\nu_{n+2}^2\}$. So $\bar{\eta}_{n+1} \circ \pi_{n+8}(M^{n+3}) = \{\eta_{n+1}\nu_{n+2}^2\} = 0$, and hence $\eta_n\bar{\eta}_{n+1} \circ \pi_{n+8}(M^{n+3}) = 0$. Thus, the indeterminacy is trivial. This establishes the proof. □

Although the following result is directly obtained from [30, Table 2], we show:

Theorem 1.34. $[\iota_n, \eta_n\varepsilon_{n+1}] = 0$ *if* $n \equiv 1 \pmod 8$ *and* $n \geq 9$.

Proof. For $n = 9$, the assertion is obtained in [38, p. 336]. By [37, p. 161] and Lemma 1.22(2), we get that

$$\pi_{n+3}(SO(n)) = 0$$

and

$$\pi_{n+4}(SO(n)) = \{[\nu_{n-2}]_n\nu_{n+1}\} \cong \mathbb{Z}_2.$$

We consider the exact sequence $(\mathcal{SO}^n_{n+1})$:

$$0 \longrightarrow \pi_{n+2}(\mathbb{S}^n) \xrightarrow{\Delta} \pi_{n+1}(SO(n)) \xrightarrow{i_*} \pi_{n+1}(SO(n+1)) \longrightarrow 0,$$

where $\pi_{n+1}(SO(n)) \cong \mathbb{Z}_8$ and $\pi_{n+1}(SO(n+1)) = \{\tau'_{n+1}\} \cong \mathbb{Z}_4$ (1.88). By Lemma 1.24(2), $i_n(\mathbb{R})\bar{\tau}'_{n-1}$ becomes a generator of $\pi_{n+1}(SO(n))$ and we have $4i_n(\mathbb{R})\bar{\tau}'_{n-1} = \Delta(\eta_n^2)$. Hence, we obtain $\Delta\eta_n \circ \eta_n\bar{\eta}_{n+1} = 0$ and we can define a Toda bracket $\{\Delta\eta_n, \eta_n\bar{\eta}_{n+1}, \tilde{\eta}_{n+2}\}_{n-5} \subseteq \pi_{n+5}(SO(n))$. By [85, Propositions 1.2:0 and 1.6 (the second formula)] and the relation $2(\eta_5\bar{\eta}_6) = 0$, we obtain

$$\begin{aligned} 2\{\Delta\eta_n, \eta_n\bar{\eta}_{n+1}, \tilde{\eta}_{n+2}\}_{n-5} &= \{\Delta\eta_n, E^{n-5}(2(\eta_5\bar{\eta}_6)), E^{n-5}\tilde{\eta}_7\}_{n-5} \\ &= \Delta\eta_n \circ E^{n-5}\pi^5_{10} + [M^{n+4}, SO(n)] \circ \tilde{\eta}_{n+3}. \end{aligned}$$

Since $E^{n-5}\pi^5_{10} = \{E^{n-5}(\nu_5\eta_8^2)\} = 0$, we have $\Delta\eta_n \circ E^{n-5}\pi^5_{10} = 0$. By the fact that $\pi_{n+3}(SO(n)) = 0$ and the relation $\nu_{n+1}\eta_{n+4} = 0$, we obtain $[M^{n+4}, SO(n)] \circ \tilde{\eta}_{n+3} = \pi_{n+4}(SO(n)) \circ \eta_{n+4} = 0$. This implies

$$(*) \qquad 2\{\Delta\eta_n, \eta_n\bar{\eta}_{n+1}, \tilde{\eta}_{n+2}\}_{n-5} = 0.$$

In virtue of [12, 14, Theorem 2] and [30, Table 1],

$$\pi_{n+4}(SO(n)) \cong \mathbb{Z}_{8d}, \text{ where } d = 2 \text{ or } 1 \text{ according as}$$
$$n \equiv 2 \pmod 8 \text{ and } n \geq 18 \text{ or } n \equiv 6 \pmod 8 \text{ and } n \geq 14 \qquad (1.103)$$

and $\pi_{n+5}(SO(n)) \cong \mathbb{Z}_{16} \oplus \mathbb{Z}_2$. By the use of the exact sequence $(\mathcal{SO}^n_{n+5})$, we see that the direct summand $\mathbb{Z}_2$ is generated by $\Delta(\nu_n^2)$. Then, by $(*)$, $\{\Delta\eta_n, \eta_n\bar{\eta}_{n+1}, \tilde{\eta}_{n+2}\}_{n-5}$ contains possibly $\Delta(\nu_n^2)$ (mod $8\pi_{n+5}(SO(n))$. By Lemma 1.33 and [85, Proposition 1.4],

$$\Delta(\eta_n\varepsilon_{n+1}) = \Delta\eta_n \circ \varepsilon_n \in \{\Delta\eta_n, \eta_n\bar{\eta}_{n+1}, \tilde{\eta}_{n+2}\}_{n-5} \circ \nu_{n+4}.$$

Thus, we obtain $\Delta(\eta_n\varepsilon_{n+1}) = a\Delta(\nu_n^3)$ for $a \in \{0, 1\}$.

Suppose that $[\iota_n, \eta_n\varepsilon_{n+1}] \neq 0$. Then, $[\iota_n, \eta_n\varepsilon_{n+1}] = [\iota_n, \nu_n^3]$. On the other hand, by [72, Proposition 4.2], $[\iota_n, \eta_n\varepsilon_{n+1}] = b[\iota_n, \eta_n^2\sigma_{n+2}]$ for $b \in \{0, 1\}$. The assumption induces the equality $[\iota_n, \eta_n\varepsilon_{n+1}] = [\iota_n, \eta_n^2\sigma_{n+2}]$. Then, we have $[\iota_n, \eta_n\varepsilon_{n+1}] = [\iota_n, \nu_n^3] + [\iota_n, \eta_n^2\sigma_{n+2}] = 2[\iota_n, \eta_n\varepsilon_{n+1}] = 0$. This completes the proof. □

Since $\pi_{4n}(SO(4n)) \cong (\mathbb{Z}_2)^3$ or $(\mathbb{Z}_2)^2$, if $n \geq 2$ [37, p. 161], we obtain

$$\sharp\tau'_{4n} = 2, \text{ if } n \geq 2. \tag{1.104}$$

The results below have been proved in [66, Proposition 2.5]. Nevertheless, we sketch proofs some of them.

Lemma 1.35. 1. *If* $n \equiv 0, 1 \pmod 4$ *and* $n \geq 8$ *then* $[\iota_n, \alpha] \neq 0$ *for* $\alpha = \varepsilon_n, \bar{\nu}_n, \eta_n\sigma_{n+1}$ *and* μ_n.
2. *If* $n \equiv 0 \pmod 4$ *and* $n \geq 8$ *then* $[\iota_n, \eta_n\mu_{n+1}] \neq 0$.

Proof. 1. We show $[\iota_n, \varepsilon_n] \neq 0$. Let $n \equiv 0 \pmod 4$ and $n \geq 8$. By [85, Proposition 11.10(i)], there exists an element $\beta \in \pi_{2n+6}^{n-1}$ such that $E\beta = [\iota_n, \varepsilon_n]$ and $H\beta = \eta_{2n-3}\varepsilon_{2n-2}$. Suppose that $[\iota_n, \varepsilon_n] = 0$. Then, by $(\mathcal{PE}_{2n+6}^{n-1})$, we have $\beta \in P\pi_{2n+8}^{2n-1}$. This induces a contradictory relation $\eta_{2n-3}\varepsilon_{2n-2} = 0$, and hence $[\iota_n, \varepsilon_n] \neq 0$.

Next, consider the case $n \equiv 1 \pmod 4$ and $n \geq 9$. Then, by (1.90), $[\iota_n, \varepsilon_n] = E(\tau_{n-1}\varepsilon_{2n-2})$ and $H(\tau_{n-1}\varepsilon_{2n-2}) = \eta_{2n-3}\varepsilon_{2n-2}$. Suppose that $[\iota_n, \varepsilon_n] = 0$. Then, $(\mathcal{PE}_{2n+6}^{n-1})$, (1.76) and (1.90) lead to a contradictory relation $\eta_{2n-3}\varepsilon_{2n-2} = 0$, and so $[\iota_n, \varepsilon_n] \neq 0$. For other α's, the argument goes ahead similarly.

2. Let $n \equiv 0 \pmod 4$ and $n \geq 8$. First, notice that [85, Proposition 11.10(i)] leads to $E\beta = [\iota_n, \eta_n\mu_{n+1}]$ for some $\beta \in \pi_{2n+8}^{n-1}$ with $H(\beta) = \eta_{2n-2}^2\mu_{2n-2}$. But, (1.64)(2) implies $H(\beta) \neq 0$. For $E\beta = 0$, by $(\mathcal{PE}_{2n+8}^{n-1})$, we get $\beta \in P\pi_{2n+10}^{2n-1} = \{[\iota_{n-1}, \iota_{n-1}] \circ \zeta_{2n-3}\}$. Consequently, the relation (1.5) yields a contraction $0 \neq H(\beta) \in \{H([\iota_{n-1}, \iota_{n-1}] \circ \zeta_{2n-3})\} = 0$ and the proof is complete.

□

By (1.6) and Lemma 1.35(1), $\Delta\colon \pi_{n+8}(\mathbb{S}^n) \to \pi_{n+7}(SO(n))$ is a monomorphism, if $n \equiv 0, 1 \pmod 4$ and $n \geq 12$. Then, by (SO_{n+8}^n), we obtain the homotopy exact sequence

$$\pi_{n+9}(\mathbb{S}^n) \xrightarrow{\Delta} \pi_{n+8}(SO(n)) \xrightarrow{i_*} \pi_{n+8}(SO(n+1)) \longrightarrow 0,$$
$$\text{if } n \equiv 0, 1 \pmod 4 \text{ and } n \geq 12. \tag{1.105}$$

We follow [85, Chap. X, (i)] to choose $\bar{\varepsilon}_3 \in \{\varepsilon_3, 2\iota_{11}, \nu_{11}^2\}_6$. In view of [85, Lemma 12.1], there exists an element $\zeta' \in \pi_{22}^6$ such that $H(\zeta') \equiv \zeta_{11} \pmod{2\zeta_{11}}$ and

$$E\zeta' = \sigma'\eta_{14}\varepsilon_{15} \text{ [85, (12.4)]}. \tag{1.106}$$

Further, (1.82)(1) and [85, Lemma 12.10] lead to:

$$\begin{aligned}&(1)\ \nu_5\sigma_8\nu_{12}^2 = \eta_5\bar{\varepsilon}_6,\\ &(2)\ \sigma'\nu_{14}^3 = \eta_7\bar{\varepsilon}_8.\end{aligned} \tag{1.107}$$

Then, (1.22), (1.69)(3), (1.106), (1.107)(2), and [85, Theorem 12.6] yield $[\iota_8, \eta_8^2\sigma_{10}] = (E\sigma')(\eta_{15}\varepsilon_{16} + \nu_{15}^3) = \eta_8\bar{\varepsilon}_9 + E^2\zeta' \neq 0$.

By (1.18), (1.22) and (1.70), we have $[\iota_9, \eta_9^2\sigma_{11}] = (\eta_9^2\sigma_{11} + \sigma_9\eta_{16}^2)\circ(\eta_{18}\sigma_{19}) = 0$. Next:

$$\begin{aligned}&(1)\ \eta_6\kappa_7 = \bar{\varepsilon}_6 \text{ [85, (10.23)]},\\ &(2)\ \kappa_7\eta_{21} = \bar{\varepsilon}_7 + \sigma'\bar{\nu}_{14} \text{ [75, Proposition 2.6(4)]}\end{aligned} \tag{1.108}$$

and

$$\begin{aligned}&(1)\ 2\kappa_7 = \bar{\nu}_7\nu_{15}^2 \text{ [53, (15.5)]},\\ &(2)\ 2\kappa_9 = \bar{\nu}_9\nu_{17}^2 = \eta_9\sigma_{10}\nu_{17}^2 \neq 0 \text{ [85, Theorem 10.3] and (1.125)}.\end{aligned} \tag{1.109}$$

The formula (1.16) and [44, Theorem C] yield

$$\sharp[\iota_n, \eta_n^2\sigma_{n+2}] = \begin{cases} 1, & \text{if } n \equiv 2, 3 \pmod 4 \text{ and } n \geq 6;\\ 2, & \text{if } n \equiv 0 \pmod 4 \text{ and } n \geq 8\end{cases} \tag{1.110}$$

and

$$\sharp[\iota_n, \eta_n^2\sigma_{n+2}] = 2, \text{ if } n \equiv 1 \pmod 8 \text{ and } n \geq 17. \tag{1.111}$$

Now, we conclude

Proposition 1.36. *$[\iota_n, \nu_n^3] = 0$ if $n \equiv 5 \pmod 8$ and $[\iota_n, \eta_n\varepsilon_{n+1}] = [\iota_n, \eta_n^2\sigma_{n+2}] = 0$ provided $n \equiv 5 \pmod 8$ and $n \geq 13$ unless $n \equiv 53 \pmod{64}$.*

Proof. By (1.54) and Lemma 1.20(1), $\Delta(\nu_n^3) = 0$ if $n \equiv 5 \pmod 8$. So, the first assertion holds. In light of [49, (7.9)], the second assertion holds for $n = 13$. Let $n \equiv 5 \pmod 8$ and $n \geq 21$. We consider the exact sequence (1.105). By [12, 14, Theorem 2] and [30, Table 1], we see that

$$\pi_{n+8}(SO(n+1)) \cong \begin{cases} \mathbb{Z}_4 \oplus \mathbb{Z}_2, & \text{if } n \equiv 5 \pmod{32} \text{ and } n \geq 37;\\ (\mathbb{Z}_4)^2, & \text{if } n \equiv 21 \pmod{32};\\ \mathbb{Z}_4, & \text{if } n \equiv 13 \pmod{16}\end{cases}$$

and

$$\pi_{n+8}(SO(n)) \cong \begin{cases} \mathbb{Z}_4 \oplus (\mathbb{Z}_2)^2, & \text{if } n \equiv 5 \pmod{32} \text{ and } n \geq 37;\\ (\mathbb{Z}_4)^2 \oplus \mathbb{Z}_2, & \text{if } n \equiv 21 \pmod{64};\\ \mathbb{Z}_8 \oplus \mathbb{Z}_4 \oplus \mathbb{Z}_2, & \text{if } n \equiv 53 \pmod{64};\\ \mathbb{Z}_4 \oplus \mathbb{Z}_2, & \text{if } n \equiv 13 \pmod{16}.\end{cases}$$

By (1.54) and (1.89), we obtain

$$\Delta(\eta_n^2\sigma_{n+2}) = 4i_n(\mathbb{R})\bar{\tau}'_{n-1}\sigma_{n+1}$$

and hence

$$\Delta(\eta_n^2\sigma_{n+2}) = \begin{cases} 0, & \text{if } n \not\equiv 53 \pmod{64}; \\ 4i_n(\mathbb{R})\bar{\tau}'_{n-1}\sigma_{n+1} \neq 0, & \text{if } n \equiv 53 \pmod{64}. \end{cases}$$

This leads to the second assertion and the proof is complete. □

Next, we show the following:

Lemma 1.37. *Let* $n \equiv 1 \pmod 4$ *and* $n \geq 5$. *Then,* $E(\bar{\tau}_{2n-2}\nu_{4n-2}^2) = [\iota_{2n-1}, \bar{\nu}_{2n-1}]$ *if and only if* $[\iota_{2n+1}, \nu_{2n+1}^2] = 0$.

Proof. By (1.91), $E^3(\bar{\tau}_{2n-2}\nu_{4n-2}^2) = [\iota_{2n+1}, \nu_{2n+1}^2]$ and this implies the necessary condition.

Suppose that $[\iota_{2n+1}, \nu_{2n+1}^2] = 0$. Then, by $(\mathcal{PE}_{4n+6}^{2n})$,

$$\pi_{4n+8}^{4n+1} \xrightarrow{P} \pi_{4n+6}^{2n} \xrightarrow{E} \pi_{4n+7}^{2n+1},$$

$E^2(\bar{\tau}_{2n-2}\nu_{4n-2}^2) \in P\pi_{4n+8}^{4n+1} \cong \mathbb{Z}_{16}$. We can set $E^2(\bar{\tau}_{2n-2}\nu_{4n-2}^2) = 8xP(\sigma_{4n+1})$ for $x \in \{0, 1\}$.

Apply [85, Proposition 11.11(ii)] to the case $\alpha = 8\sigma_{4n-6}$, then there exists an element $\beta \in \pi_{4n+4}^{2n-2}$ such that

$$P(8\sigma_{4n+1}) = E^2\beta \quad \text{and} \quad H(\beta) \in \{\eta_{4n-5}, 2\iota_{4n-4}, 8\sigma_{4n-4}\}_2.$$

By [85, Lemma 6.5, Theorem 7.1] and (1.68),

$$\mu_{4n-5} \in \{\eta_{4n-5}, 2\iota_{4n-4}, 8\sigma_{4n-4}\}_2 \bmod \eta_{4n-5} \circ E^2\pi_{4n+2}^{4n-6} = \{\nu_{4n-5}^3, \eta_{4n-5}\varepsilon_{4n-4}\}.$$

So we obtain

$$H(\beta) = \mu_{4n-5} + y\nu_{4n-5}^3 + z\eta_{4n-5}\varepsilon_{4n-4} \ (y, z \in \{0, 1\}).$$

By using $(\mathcal{PE}_{4n+5}^{2n-1})$ and the assumption,

$$E(\bar{\tau}_{2n-2}\nu_{4n-2}^2) - xE\beta \in P\pi_{4n+7}^{4n-1} = \{P(\bar{\nu}_{4n-1}), P(\varepsilon_{4n-1})\}.$$

By Lemma 1.24(1), $P(\bar{\nu}_{4n-1}) = E(\tau_{2n-2}\bar{\nu}_{4n-4})$ and $P(\varepsilon_{4n-1}) = E(\tau_{2n-2}\varepsilon_{4n-4})$. Then, by using $(\mathcal{PE}_{4n+4}^{2n-2})$,

$$\bar{\tau}_{2n-2}\nu_{4n-2}^2 - x\beta - a\tau_{2n-2}\bar{\nu}_{4n-4} - b\tau_{2n-2}\varepsilon_{4n-4} \in P\pi_{4n+6}^{4n-3} \ (a, b \in \{0, 1\}).$$

By applying $H: \pi^{2n-2}_{4n+5} \to \pi^{4n-5}_{4n+5}$ to the equation, by the use of (1.90), (1.91) and (1.68), we obtain

$$\nu^3_{4n-5} - x(\mu_{4n-5} + y\nu^3_{4n-5} + z\eta_{4n-5}\varepsilon_{4n-4}) = a\nu^3_{4n-5} + b\eta_{4n-5}\varepsilon_{4n-4}.$$

Since $\mu_{4n-5}, \nu^3_{4n-5}, \eta_{4n-5}\varepsilon_{4n-4}$ generate π^{4n-5}_{4n+4} independently, we have $x = 0$, $a = 1$ and $b = 0$. Hence, $E(\bar{\tau}_{2n-2}\nu^2_{4n-2}) = E(\tau_{2n-2}\bar{\nu}_{4n-4})$. This completes the proof. □

Since $\nu_n\eta_{n+3} = 0$ (1.33) and $\bar{\nu}_n\eta_{n+8} = \nu^3_n$ (1.68) for $n \geq 6$, Lemma 1.37 implies

Corollary 1.38. *If* $[\iota_{8n+3}, \nu^2_{8n+3}] = 0$, *then* $[\iota_{8n+1}, \nu^3_{8n+1}] = 0$.

Now, we show

III. $[\iota_n, \nu^2_n] = 0$ **if** $n = 2^i - 5$ $(i \geq 4)$**.**

We recall the Mahowald element $\eta'_i \in \pi^S_{2^i}(\mathbb{S}^0)$ [45, Theorem 1] for $i \geq 3$. We set $\eta'_{i-1,m} = \eta'_{i-1}$ on $\mathbb{S}^m$ for $m = 2^{i-1} - 2$ with $i \geq 4$, that is, $\eta'_{i-1,m} \in \pi_{2^{i-1}+m}(\mathbb{S}^m)$. It satisfies the relation $H(\eta'_{i-1,m}) = \nu_{2m-1}$. Then, the assertion follows directly from [7, Proposition] taking $\alpha = \beta = \eta'_{i-1,m}$.

Finally, we show

IV. $[\iota_n, \nu^2_n] \neq 0$ **if** $n \equiv 3 \pmod 8$ **and** $n \geq 19$ **unless** $n = 2^i - 5$**.**

By III and Corollary 1.38, we obtain

$$[\iota_n, \nu^3_n] = 0, \quad \text{if} \quad n = 2^i - 7 \ (i \geq 4).$$

Hence, from Theorem 1.34 and the relation $\eta^2_n\sigma_{n+2} = \nu^3_n + \eta_n\varepsilon_{n+1}$,

$$[\iota_n, \eta^2_n\sigma_{n+2}] = 0, \text{ if } n = 2^i - 7 \ (i \geq 4).$$

Let $n \equiv 1 \pmod 8$ and $n \geq 17$. Considering the exact sequence (1.105), in virtue of [12, 14, Theorem 2] and [30, Table 1], we obtain

$$\pi_{n+8}(SO(n)) \cong \mathbb{Z}_2 \oplus \mathbb{Z}_2 \oplus \mathbb{Z}_8 \quad \text{and} \quad \pi_{n+8}(SO(n+1)) \cong \mathbb{Z}_2 \oplus \mathbb{Z}_4.$$

By (1.92) and (1.111), we get the relation

$$4E(\bar{\tau}_{n-1}\sigma_{2n}) = [\iota_n, \eta^2_n\sigma_{n+2}] \neq 0.$$

Hence, by (1.111) and Theorem 1.34, we obtain

$$[\iota_n, \nu^3_n] = [\iota_n, \eta^2_n\sigma_{n+2}] \neq 0, \text{ if } n \equiv 1 \pmod 8 \text{ and } n \geq 17 \text{ and } n \neq 2^i - 7.$$

Thus, by Corollary 1.38, we obtain the assertion.

We are in a position to assert that Mahowald's result [64, Table 2 for $\eta^2\rho_1$] should be stated as follows.

Theorem 1.39. *Let* $n \equiv 1 \pmod 8$ *and* $n \geq 9$. *Then* $[\iota_n, \eta^2_n\sigma_{n+2}] \neq 0$ *if and only if* $n \neq 2^i - 7$.

1.5 Proof of $[\iota_{16s+7}, \sigma_{16s+7}] \neq 0$ for $s \geq 1$

We present a proof of the first part of Theorem 1.17: $[\iota_{16s+7}, \sigma_{16s+7}] \neq 0$ for $s \geq 1$. By [37, p. 161], $\pi_{n+4}(SO(n+k)) \cong \mathbb{Z} \oplus \mathbb{Z}_2$ for $k = 1, 2$ if $n \equiv 7 \pmod 8$. And, by $(\mathcal{SO}^{n+2}_{n+4})$, the direct summand $\mathbb{Z}_2$ of $\pi_{n+4}(SO(n+2))$ is generated by $\Delta\nu_{n+2}$. So, the non-triviality of $[\nu_n]\eta_{n+3} \in \pi_{n+4}(SO(n+1))$ induces the relation $i_{n+2}(\mathbb{R})_*([\nu_n]\eta_{n+3}) = \Delta\nu_{n+2}$. Because of the fact that $[\iota_{n+2}, \nu^2_{n+2}] \neq 0$, this induces a contradictory relation $0 = \Delta\nu^2_{n+2} \neq 0$. Hence, we obtain

$$[\nu_n]\eta_{n+3} = 0, \text{ if } n \equiv 7 \pmod 8.$$

Next, by [37, p. 161],

$$\{[\nu_n], \eta_{n+3}, 2\iota_{n+4}\} \subseteq \pi_{n+5}(SO(n+1)) = 0, \text{ if } n \equiv 7 \pmod 8.$$

So, by (1.93), we have $[\nu_n]\bar{\eta}_{n+3} \in \{[\nu_n], \eta_{n+3}, 2\iota_{n+4}\} \circ p_{n+5} = 0$ and hence we can define a lift of κ_n for $n \equiv 7 \pmod 8$, as follows:

$$[\kappa_n] \in \{[\nu_n], \bar{\eta}_{n+3}, \tilde{\bar{\nu}}_{n+4}\} \subseteq \pi_{n+14}(SO(n+1)) \text{ for } n \equiv 7 \pmod 8.$$

Let $n \equiv 7 \pmod 8$ and $n \geq 15$. By the use of $(\mathcal{SO}^{n-k}_{n-4})$ for $k = 3, 4$, $(\mathcal{SO}^{n-l}_{n-3})$ for $l = 2, 3, 5$, $(\mathcal{SO}^{n-m}_{n-2})$ for $2 \leq m \leq 5$ and [37, p. 161], we obtain

$$\pi_{n-4}(SO(n-4)) = \{\beta\} \cong \mathbb{Z};\ \pi_{n-4}(SO(n-3)) = \{i_{n-3}(\mathbb{R})\beta, \Delta\iota_{n-3}\} \cong (\mathbb{Z})^2;$$

$$\pi_{n-3}(SO(n-4)) = \{[\eta^2_{n-5}]\} \cong \mathbb{Z}_2;\ \pi_{n-3}(SO(n-3)) = \{[\eta_{n-4}], \Delta\eta_{n-3}\} \cong (\mathbb{Z}_2)^2;$$

$$\pi_{n-2}(SO(n-4)) = \{[\eta^2_{n-5}]\eta_{n-3}, \Delta\nu_{n-4}\} \cong (\mathbb{Z}_2)^2;$$

$$\pi_{n-2}(SO(n-3)) = \{[\eta_{n-4}]\eta_{n-3}, \Delta\eta^2_{n-3}\} \cong (\mathbb{Z}_2)^2;\ \pi_{n-2}(SO(n-2)) = \{\Delta\eta_{n-2}\} \cong \mathbb{Z}_2,$$

where β is a generator of $\pi_{n-4}(SO(n-4))$ and

$$\Delta\eta_{n-3} = [\eta^2_{n-5}]_{n-3}. \qquad (1.112)$$

We need

$$\{p_n(\mathbb{R}), i_n(\mathbb{R}), \Delta\iota_{n-1}\} \ni \iota_{n-1} \pmod{2\iota_{n-1}} \text{ for } n \geq 9. \qquad (1.113)$$

By the same reason as (1.52), we obtain $\Delta(\bar{\eta}_3) = 0 \in [M^4, SO(3)]$.

Next, by Lemma 1.18(1) and (1.93), we obtain

$$\Delta(\bar{\eta}_{n-4}) = \Delta\iota_{n-4} \circ \bar{\eta}_{n-5} \in -\{\Delta\iota_{n-4}, \eta_{n-5}, 2\iota_{n-4}\} \circ p_{n-3} = 0.$$

So, $\bar{\eta}_{n-4}$ is lifted to $[\bar{\eta}_{n-4}] \in [M^{n-2}, SO(n-3)]$ for $n \equiv 7 \pmod 8$. We set $[\bar{\eta}_{n-4}] \circ i_{n-2} = [\eta_{n-4}]$, which is a lift of η_{n-4}. By (1.112) and (1.113), we get

$$[\eta_{n-4}] \in \{i_{n-3}(\mathbb{R}), \Delta\iota_{n-4}, \eta_{n-5}\} \pmod{i_{n-3}(\mathbb{R}) \circ \pi_{n-3}(SO(n-4)) + \pi_{n-4}(SO(n-3)) \circ \eta_{n-4} = \{\Delta\eta_{n-3}\}} \text{ for } n \equiv 7 \pmod 8 \text{ and } n \geq 15. \tag{1.114}$$

Now, the map $\eta^*_{n-3} : \pi_{n-3}(SO(n-3)) \to \pi_{n-2}(SO(n-3))$ is an isomorphism. Hence, Lemma 1.25 and the relation $[\bar{\eta}_{n-4}] \circ i_{n-2} = [\eta_{n-4}]$ yield

$$\overline{[\eta_{n-4}]} \equiv [\bar{\eta}_{n-4}] \pmod{\pi_{n-2}(SO(n-3)) \circ p_{n-2} = 2[M^{n-2}, SO(n-3)]}. \tag{1.115}$$

We show

Lemma 1.40. *Let* $n \equiv 7 \pmod 8$ *and* $n \geq 15$. *Then,*

1. $\overline{[\eta_{n-4}]} \in \{i_{n-3}(\mathbb{R}), \Delta\iota_{n-4}, \bar{\eta}_{n-5}\} \pmod{\{\Delta(\bar{\eta}_{n-3})\} + K}$, *where* $K = i_{n-3}(\mathbb{R})_*[M^{n-2}, SO(n-4)] + \pi_{n-4}(SO(n-3)) \circ \bar{\eta}_{n-4}$*;*
2. $i_{n-2}(\mathbb{R})_* K \subseteq \{(\Delta\eta_{n-2})p_{n-2}\}$.

Proof. By (1.40), (1.115) and (1.114), we have (1).

We see that $[M^{n-2}, SO(n-4)] = \{\overline{[\eta^2_{n-5}]}, (\Delta\nu_{n-4})p_{n-2}\} \cong \mathbb{Z}_4 \oplus \mathbb{Z}_2$, where $\overline{[\eta^2_{n-5}]}$ is an extension of $[\eta^2_{n-5}]$ and $2\overline{[\eta^2_{n-5}]} = [\eta^2_{n-5}]\eta_{n-3}p_{n-2}$. Hence, by (1.112),

$$i_{n-4,n-2*}\overline{[\eta^2_{n-5}]} \in i_{n-2}(\mathbb{R}) \circ \{\Delta\eta_{n-3}, 2\iota_{n-3}, p_{n-3}\} =$$

$$-\{i_{n-2}(\mathbb{R}), \Delta\eta_{n-3}, 2\iota_{n-3}\} \circ p_{n-2}.$$

Since $\{i_{n-2}(\mathbb{R}), \Delta\eta_{n-3}, 2\iota_{n-3}\} \subseteq \pi_{n-2}(SO(n-2)) = \{\Delta\eta_{n-2}\}$, we have $i_{n-4,n-2*}[M^{n-2}, SO(n-4)] \subseteq \{(\Delta\eta_{n-2})p_{n-2}\}$.

From the relation $p_{n-3}(\mathbb{R})\beta = 0$, we obtain $\beta\eta_{n-4} = 0 \in \pi_{n-3}(SO(n-4))$. Then, by (1.93), we have $\beta\bar{\eta}_{n-4} \in \{\beta, \eta_{n-4}, 2\iota_{n-3}\} \circ p_{n-2} \subseteq \pi_{n-2}(SO(n-2)) \circ p_{n-2}$. Hence, we obtain $i_{n-2}(\mathbb{R})_*(\pi_{n-4}(SO(n-3)) \circ \bar{\eta}_{n-4}) \subseteq \{(\Delta\eta_{n-2})p_{n-2}\}$. This leads to (2) and completes the proof. □

We show

Lemma 1.41. $[\kappa_{n-8}]_{n-1} = \Delta\bar{\nu}_{n-1}$ *if* $n \equiv 7 \pmod 8$ *and* $n \geq 15$.

Proof. By the use of $(\mathcal{SO}^{n-7+k}_{n-5})$ for $0 \leq k \leq 3$ and [37, p. 161], we have $[\nu_{n-8}]_{n-4} = \Delta\iota_{n-4}$, and so

$$[\kappa_{n-8}]_{n-1} \in (i_{n-4,n-1})_*\{\Delta\iota_{n-4}, \bar{\eta}_{n-5}, \tilde{\bar{\nu}}_{n-4}\}.$$

By (1.115) and Lemma 1.40, we obtain

$$i_{n-3}(\mathbb{R})_*\{\Delta\iota_{n-4}, \bar{\eta}_{n-5}, \bar{\bar{\nu}}_{n-4}\} = -\{i_{n-3}(\mathbb{R}), \Delta\iota_{n-4}, \bar{\eta}_{n-5}\} \circ \bar{\bar{\nu}}_{n-3}$$

$$\equiv \overline{[\eta_{n-4}]} \circ \bar{\bar{\nu}}_{n-3} \in \{[\eta_{n-4}], 2\iota_{n-3}, \bar{\nu}_{n-3}\}$$

$$(\text{mod } [\eta_{n-4}] \circ \pi_{n+6}(\mathbb{S}^{n-3}) + \pi_{n-2}(SO(n-3)) \circ \bar{\nu}_{n-2} + K \circ \bar{\bar{\nu}}_{n-3}).$$

By Lemma 1.40 and (1.74), $i_{n-2}(\mathbb{R})_*(K \circ \bar{\bar{\nu}}_{n-3}) \subseteq \{\Delta\eta_{n-2}\} \circ \bar{\nu}_{n-3} = \{\Delta\nu^3_{n-2}\} = 0$. From the relation $[\eta_{n-4}]_{n-2} = \Delta\iota_{n-2}$, we see that

$$[\kappa_{n-8}]_{n-2} \in \{\Delta\iota_{n-2}, 2\iota_{n-3}, \bar{\nu}_{n-3}\} \ (\text{mod } \Delta\pi_{n+7}(\mathbb{S}^{n-2}))$$

and

$$[\kappa_{n-8}]_{n-1} \in -i_{n-1}(\mathbb{R}) \circ \{\Delta\iota_{n-2}, 2\iota_{n-3}, \bar{\nu}_{n-3}\}$$
$$= \{i_{n-1}(\mathbb{R}), \Delta\iota_{n-2}, 2\iota_{n-3}\} \circ \bar{\nu}_{n-2}.$$

Since $\{i_{n-1}(\mathbb{R}), \Delta\iota_{n-2}, 2\iota_{n-3}\} \equiv \Delta\iota_{n-1}$ (mod $2\Delta\iota_{n-1}$) by (1.113), we have

$$\{i_{n-1}(\mathbb{R}), \Delta\iota_{n-2}, 2\iota_{n-3}\} \circ \bar{\nu}_{n-2} = \Delta\bar{\nu}_{n-1}.$$

This completes the proof. □

Hereafter, we fix $n = 16s + 7 \geq 23$. Suppose that $E^7(\gamma\sigma_{2n-8}) = [\iota_n, \sigma_n] = 0$, where γ is the element in (1.43). Then, by $(\mathcal{PE}^{n-1}_{2n+5})$ and Lemma 1.41, $E^6(\gamma\sigma_{2n-8}) \in \{[\iota_{n-1}, \bar{\nu}_{n-1}] = E^6 J[\kappa_{n-7}], [\iota_{n-1}, \eta_{n-1}\sigma_n]\}$.

By [69, p. 382: Table], there exists an element $\delta \in \pi^{n-8}_{2n-10}$ such that

$$[\iota_{n-1}, \eta_{n-1}] = E^7\delta \text{ and } H\delta = \sigma_{2n-17} \tag{1.116}$$

and so, $[\iota_{n-1}, \eta_{n-1}\sigma_n]$ desuspends until we reach seven dimensions. Hence, in the sequel argument, it suffices to consider $E^6(\gamma\sigma_{2n-8}) = aE^6 J[\kappa_{n-7}]$ for $a \in \{0, 1\}$. By $(\mathcal{PE}^{n-2}_{2n+4})$, we have

$$E^5(\gamma\sigma_{2n-8} - aJ[\kappa_{n-7}]) \in P\pi^{2n-3}_{2n+6}.$$

By Lemma 1.35(1) and Proposition 1.36, $P\mu_{2n-3} \neq 0$ and $P(\nu^3_{2n-3}) = 0$. By [69, p. 383: Table], $[\iota_{n-2}, \eta^2_{n-2}]$ and $[\iota_{n-2}, \eta^2_{n-2}\sigma_n]$ desuspend until 7 dimensions. Hence, for $x \in \{0, 1\}$, we have

$$E^5(\gamma\sigma_{2n-8} - aJ[\kappa_{n-7}]) = xP\mu_{2n-3}.$$

By [85, Proposition 11.10(ii)], there exists an element $\beta \in \pi^{n-3}_{2n+3}$ such that $P\mu_{2n-3} = E\beta$ and $H\beta = \eta_{2n-7}\mu_{2n-6}$. Then, by $(\mathcal{PE}^{n-3}_{2n+3})$, we have

$$E^4(\gamma\sigma_{2n-8} - aJ[\kappa_{n-7}]) - x\beta \in P\pi^{2n-5}_{2n+5}.$$

This induces the relation $x\eta_{2n-7}\mu_{2n-6} = 0$. Hence, $x = 0$ and we can set

$$E^4(\gamma\sigma_{2n-8} - aJ[\kappa_{n-7}]) = yP(\eta_{2n-5}\mu_{2n-4}) \text{ for } y \in \{0, 1\}.$$

By [85, Proposition 11.10(i)], there exists an element $\beta' \in \pi^{n-4}_{2n+2}$ such that $P(\eta_{2n-5}\mu_{2n-4}) = E\beta'$ and $H\beta' = \eta^2_{2n-9}\mu_{2n-7}$. So, we have

$$E^3(\gamma\sigma_{2n-8} - aJ[\kappa_{n-7}]) - y\beta' \in P\pi^{2n-7}_{2n+4}.$$

This leads to the relation $y\eta^2_{2n-9}\mu_{2n-7} = 0$, and hence $y = 0$. Therefore, by (1.91), we obtain

$$E^3(\gamma\sigma_{2n-8} - aJ[\kappa_{n-7}] - b\bar{\tau}_{n-7}\zeta_{2n-12}) = 0 \ (b \in \{0, 1\}).$$

By $(\mathcal{PE}^{n-5-k}_{2n+1-k})$ for $k = 0, 1$ and 2, we have

$$E^2(\gamma\sigma_{2n-8} - aJ[\kappa_{n-7}] - b\bar{\tau}_{n-7}\zeta_{2n-12}) \in P\pi^{2n-9}_{2n+3} = 0$$

$$E(\gamma\sigma_{2n-8} - aJ[\kappa_{n-7}] - b\bar{\tau}_{n-7}\zeta_{2n-12}) \in P\pi^{2n-11}_{2n+2} = 0$$

and

$$\gamma\sigma_{2n-8} - aJ[\kappa_{n-7}] - b\bar{\tau}_{n-7}\zeta_{2n-12} \in P\pi^{2n-13}_{2n+1}.$$

By (1.91) and [85, Lemma 9.2, Theorem 10.3], $H(\bar{\tau}_{n-7}\zeta_{2n-12}) = \pm(\frac{n-3}{4})\nu_{2n-15}\zeta_{2n-12} = \pm 2(n-3)\sigma^2_{2n-15} = 0$. Then, the last relation induces the contradictory relation $\sigma^2_{2n-15} = a\kappa_{2n-15}$. Thus, we obtain the non-triviality of $[\iota_n, \sigma_n]$ if $n \equiv 7 \pmod{16}$ and $n \geq 23$.

By Lemma 1.41, we have $[\iota_n, \bar{\nu}_n] = E^6J[\kappa_{n-7}]$ if $n \equiv 6 \pmod 8$ and $n \geq 14$. By the parallel arguments to the above, we obtain

Corollary 1.42. *$[\iota_n, \bar{\nu}_n] \neq 0$, if $n \equiv 6 \pmod 8$ and $n \geq 14$.*

1.6 Gottlieb Groups of Spheres with Stems for $8 \leq k \leq 13$

By [85, Theorems 7.1, 7.4, 7.6, p. 186: Table], $\pi_{n+8}(\mathbb{S}^n) = \{\varepsilon_n\} \cong \mathbb{Z}_2$ for $n = 4, 5$ and $[\iota_4, \varepsilon_4] = (E\nu')\varepsilon_7 \neq 0$, $[\iota_5, \varepsilon_5] = \nu_5\eta_8\varepsilon_9 \neq 0$.

We recall $\pi_{14}(\mathbb{S}^6) = \{\bar{\nu}_6, \varepsilon_6, [\iota_6, \alpha_1(6)]\} \cong \mathbb{Z}_{24} \oplus \mathbb{Z}_2$. By [85, (7.27)],

$$[\iota_6, \bar{\nu}_6] = [\iota_6, \varepsilon_6] = 0. \tag{1.117}$$

So, we obtain $G_{14}(\mathbb{S}^6; 2) = \pi^6_{14}$. By Proposition 1.10(1), $G_{14}(\mathbb{S}^6; 3) = \pi_{14}(\mathbb{S}^6; 3)$. This shows $G_{14}(\mathbb{S}^6) = \pi_{14}(\mathbb{S}^6)$.

We recall $\pi_{16}(\mathbb{S}^8) = \{\sigma_8\eta_{15}, (E\sigma')\eta_{15}, \bar{\nu}_8, \varepsilon_8\} \cong (\mathbb{Z}_2)^4$ and $\pi_{17}(\mathbb{S}^9) = \{\sigma_9\eta_{16}, \bar{\nu}_9, \varepsilon_9\} \cong (\mathbb{Z}_2)^3$. We have $[\iota_8, \sigma_8\eta_{15}] = (E\sigma')\sigma_{15}\eta_{22} = (E\sigma')(\bar{\nu}_{15} + \varepsilon_{15}) = [\iota_8, \bar{\nu}_8] + [\iota_8, \varepsilon_8]$. By (1.44) and [85, Theorem 12.6], $[\iota_9, \sigma_9\eta_{16}] = \sigma_9(\nu_{16}^3 + \eta_{16}\varepsilon_{17}) \neq 0$. Hence, we obtain

$$G_{16}(\mathbb{S}^8) = \{(E\sigma')\eta_{15}, \sigma_8\eta_{15} + \bar{\nu}_8 + \varepsilon_8\} \cong (\mathbb{Z}_2)^2 \text{ and } G_{17}(\mathbb{S}^9) = \{[\iota_9, \iota_9]\} \cong \mathbb{Z}_2.$$

Hence, by Lemma 1.35, we get that

$$G_{n+8}(\mathbb{S}^n) = 0, \quad \text{if} \quad n \equiv 0, 1 \pmod 4 \text{ and } n \geq 4 \text{ unless } n = 8, 9.$$

Since $\pi_{27}(\mathbb{S}^{10}) \to \pi_{28}(\mathbb{S}^{11})$ is a monomorphism [85, (12.21)], we obtain

$$G_{18}(\mathbb{S}^{10}) = \pi_{18}(\mathbb{S}^{10}).$$

Let $n \equiv 3 \pmod 4$ and $n \geq 11$. Then, by Lemma 1.2(1) and (1.15), $[\iota_n, \eta_n\sigma_{n+1}] = 0$. In virtue of (1.6) and Example 1.19, we obtain $[\iota_n, \varepsilon_n] = 0$. Thus, as it is expected in Proposition 1.8,

$$G_{n+8}(\mathbb{S}^n) = \pi_{n+8}(\mathbb{S}^n), \quad \text{if} \quad n \equiv 3 \pmod 4.$$

By Lemma 1.35(1) and [44, Theorem C],

$$\sharp[\iota_n, \eta_n\sigma_{n+1}] = \begin{cases} 2, & \text{if } n \equiv 0, 1, 2, 4, 5 \pmod 8 \text{ and } n \geq 8 \text{ unless } n = 10; \\ 1, & \text{if } n \equiv 3 \pmod 4 \text{ and } n \geq 7. \end{cases} \tag{1.118}$$

Here we recall from [9, p. 137, Corollary 1.6] and [17, p. 48: Theorem], the following:

Theorem 1.43 (Barratt–Jones–Feder–Gitler–Lam–Mahowald). *Let β's generate the J-image in the s-stem and assume $3s - 2 \leq 2n$. Then,*

1. $[\iota_n, \beta] = 0$, *provided n and s satisfy* $3 \leq \nu_2(n + s + 2) \leq \phi(s)$;
2. $[\iota_n, \beta] \neq 0$ *provided n and s satisfy* $\nu_2(n+s+2) \geq \phi(s)+1 \geq 3$, *but* $n+s+2 \neq 2^{\phi(s)+1}$.

Here $\nu_2(m)$ is the exponent of 2 in the factorization of m and $\phi(s)$ denotes the number of integers in the closed interval $[1, s]$ which are congruent to 0, 1, 2, or 4 modulo 8.

By the use of Theorem 1.43, we obtain

$$\sharp[\iota_n, \eta_n\sigma_{n+1}] = \begin{cases} 2, & \text{if } n \equiv 22 \pmod{32} \text{ and } n \geq 54; \\ 1, & \text{if } n \equiv 14 \pmod{16} \text{ or } n \equiv 6 \pmod{32} \text{ and } n \geq 14 \end{cases} \tag{1.119}$$

and

$$\sharp[\iota_n, \eta_n^2\sigma_{n+2}] = \begin{cases} 2, & \text{if } n \equiv 53 \pmod{64} \text{ and } n \geq 117; \\ 1, & \text{if } n \equiv 13 \pmod{16},\ 5 \pmod{32} \text{ or } 21 \pmod{64} \text{ and } n \geq 13. \end{cases} \tag{1.120}$$

Now, we show

Lemma 1.44. 1. *Let* $n \equiv 2 \pmod 8$ *and* $n \geq 18$. *Then,* $\Delta\varepsilon_n = 0$.
2. *Let* $n \equiv 6 \pmod 8$ *and* $n \geq 14$. *Then,* $\Delta\varepsilon_n = \pm 2[\nu_{n-2}^2]_n\nu_{n+4}$.

Proof. Although (1) is directly obtained by [30, Table 2], we give a different proof.

Let $n \equiv 2 \pmod 4$ and $n \geq 18$. Then, by the fact that $\pi_{n+1}(SO(n)) \cong \mathbb{Z}$ [37, p. 161], we have $\tau_n'\eta_n = 0$. Then, by (1.54), (1.95), and (1.88), we obtain

$$\Delta(\eta_n\bar{\eta}_{n+1}) = 2\tau_n' \circ \bar{\eta}_n = \tau_n' \circ \eta_n^2 p_{n+2} = 0.$$

Therefore, by Lemma 1.33, we get

$$\Delta\varepsilon_n = \Delta\iota_n \circ \varepsilon_{n-1} = \Delta\iota_n \circ \{\eta_{n-1}\bar{\eta}_n, \tilde{\eta}_{n+1}, \nu_{n+3}\} = -\{\Delta\iota_n, \eta_{n-1}\bar{\eta}_n, \tilde{\eta}_{n+1}\} \circ \nu_{n+4}.$$

We have

$$\{\Delta\iota_n, \eta_{n-1}\bar{\eta}_n, \tilde{\eta}_{n+1}\} \subseteq \pi_{n+4}(SO(n)).$$

Noting the relation $4\tilde{\eta}_{n+1} = 0$, we obtain

$$4\{\Delta\iota_n, \eta_{n-1}\bar{\eta}_n, \tilde{\eta}_{n+1}\} = -\Delta\iota_n \circ \{\eta_{n-1}\bar{\eta}_n, \tilde{\eta}_{n+1}, 4\iota_{n+3}\} \subseteq -\Delta\iota_n \circ \pi_{n+4}(\mathbb{S}^{n-1}) = 0.$$

This induces $\Delta\varepsilon_n \in (2d)(\pi_{n+4}(SO(n)) \circ \nu_{n+4})$, where d is the number in (1.103). Since $4\pi_{n+7}(SO(n)) = 0$ by [12, 14, Theorem 2] and [30, Table 1], we obtain (1).

Let $n \equiv 6 \pmod 8$ and $n \geq 14$. By the exact sequences $(\mathcal{SO}_{n+4}^{n+k})$ for $k = -2, -1$ and Lemma 1.20 we get that $i_n(\mathbb{R})_*\colon \pi_{n+4}(SO(n-1)) \to \pi_{n+4}(SO(n))$ is an isomorphism and $\pi_{n+4}(SO(n-1)) = \{[\nu_{n-2}^2]\} \cong \mathbb{Z}_8$.

By [30, Table 2], $\Delta\varepsilon_n \neq 0$ for $n \equiv 6 \pmod 8$ and $n \geq 14$. Hence, (2) follows and the proof is complete. □

Now, by Lemma 1.44(1) and (1.118),

$$[\iota_n, \varepsilon_n] = 0 \text{ and } [\iota_n, \bar{\nu}_n] = [\iota_n, \eta_n\sigma_{n+1}] \neq 0, \text{ if } n \equiv 2 \pmod 8 \text{ and } n \geq 18.$$

Whence, we conclude that

$$G_{n+8}(\mathbb{S}^n) = \{\varepsilon_n\} \cong \mathbb{Z}_2, \text{ if } n \equiv 2 \pmod 8 \text{ and } n \geq 18.$$

We show $[\iota_n, \varepsilon_n] \neq 0$ if $n \equiv 22 \pmod{32}$ and $n \geq 22$. By (1.116), there exists an element $\delta \in \pi_{2n-8}^{n-7}$ such that $[\iota_n, \eta_n] = E^7\delta$ and $H\delta = \sigma_{2n-15}$. Hence, by

Lemma 1.41, $[\iota_n, \varepsilon_n] = E^6(J[\kappa_{n-7}] + E(\delta\sigma_{2n-7}))$. Suppose that $[\iota_n, \varepsilon_n] = 0$. Then, by the parallel argument to that in the proof the non-triviality of $[\iota_{n+1}, \sigma_{n+1}]$, we get a contradiction.

By [49, (7.13)], $\mathrm{Ker}\{P\colon \pi_{37}(\mathbb{S}^{29}) \to \pi_{35}(\mathbb{S}^{14})\} = \{\eta_{14}\sigma_{15}\}$ and hence $G_{22}(\mathbb{S}^{14}) = \{\eta_{14}\sigma_{15}\} \cong \mathbb{Z}_2$. By [73, p. 134: (7.29)], $\mathrm{Ker}\{P\colon \pi_{53}^{45} \to \pi_{51}^{22}\} = \{\eta_{45}\sigma_{46}\}$ and hence $G_{30}(\mathbb{S}^{22}) = \{\eta_{22}\sigma_{23}\} \cong \mathbb{Z}_2$. Thus, we have shown

Proposition 1.45. *The group* $G_{n+8}(\mathbb{S}^n)$ *is equal to the following:* 0 *if* $n \equiv 0, 1 \pmod 4$ *and* $n \geq 4$ *unless* $n = 8, 9$ *or* $n \equiv 22 \pmod{32}$ *and* $n \geq 54$*;* $\pi_{n+8}(\mathbb{S}^n)$ *if* $n = 6, 10$ *or* $n \equiv 3 \pmod 4$*;* $\{\varepsilon_n\} \cong \mathbb{Z}_2$*, if* $n \equiv 2 \pmod 8$ *and* $n \geq 18$*. Moreover,* $G_{n+8}(\mathbb{S}^n) = \{\eta_n\sigma_{n+1}\} \cong \mathbb{Z}_2$ *if* $n = 22$*,* $n \equiv 14 \pmod{16}$ *or* $n \equiv 6 \pmod{32}$ *with* $n \geq 14$*;* $G_{16}(\mathbb{S}^8) = \{(E\sigma')\eta_{15}, \sigma_8\eta_{15} + \bar{\nu}_8 + \varepsilon_8\} \cong (\mathbb{Z}_2)^2$ *and* $G_{17}(\mathbb{S}^9) = \{[\iota_9, \iota_9]\} \cong \mathbb{Z}_2$.

Recall that by [85, Chap. VI, iv)] the Toda bracket $\{\nu', 2\nu_6, \nu_9\}_3$ consists of a single element which is denoted by

$$\varepsilon' = \{\nu', 2\nu_6, \nu_9\}_3 \in \pi_{13}(\mathbb{S}^3) \text{ and } 2\varepsilon' = \eta_3^2\varepsilon_5 \text{ [85, Lemma 6.6].} \tag{1.121}$$

Then

$$\begin{matrix}(1)\ \varepsilon_3\nu_{11} = \nu'\bar{\nu}_6 \\ (2)\ \varepsilon'\eta_{13} = \nu'\varepsilon_6\end{matrix} \text{ [85, (7.12)].} \tag{1.122}$$

Notice that (1.25) and (1.122)(1) yield

$$\varepsilon_5\nu_{13} = 0. \tag{1.123}$$

Next, in view of (1.121) and (1.123), we get $\nu' E^3\varepsilon' \in \nu' \circ \{2\nu_6, \nu_9, 2\nu_{12}\}_3 = \{\nu', 2\nu_6, \nu_9\}_3 \circ 2\nu_{13} = \varepsilon' \circ 2\nu_{13} = \eta_3^2\varepsilon_5\nu_{13} = 0$. Hence,

$$\nu' E^3\varepsilon' = 0. \tag{1.124}$$

The result [85, Theorem 7.4], (1.69)(1), and (1.123) lead to

$$\eta_9\sigma_{10}\nu_{17} = \bar{\nu}_9\nu_{17} \neq 0. \tag{1.125}$$

Next, the result [85, (7.17), (7.18)], and (1.67) yield

$$(2\bar{\nu}_6)\nu_{14} = \nu_6\varepsilon_9 = \nu_6\bar{\nu}_9 = [\iota_6, \nu_6^2]. \tag{1.126}$$

Further, we recall the results:

$$E\theta' = [\iota_{12}, \eta_{12}] \text{ [85, p. 80],} \tag{1.127}$$

$$E\theta = [\iota_{13}, \iota_{13}] \text{ [85, (7.30)]} \tag{1.128}$$

and

$$\bar{\nu}_9\nu_{17} = [\iota_9, \nu_9]\ [85, (7.22)]. \tag{1.129}$$

By [85, Theorem 7.6],

$$[\iota_4, \mu_4] = (E\nu')\mu_7 \neq 0. \tag{1.130}$$

We have $[\iota_5, \mu_5] = \nu_5\eta_8\mu_9 \neq 0$ [85, Theorem 7.7].

We recall that $\pi_{15}(\mathbb{S}^6) = \pi^6_{15} = \{\nu_6^3, \mu_6, \eta_6\varepsilon_7\} \cong (\mathbb{Z}_2)^3$. Because the map $E : \pi^6_{20} \to \pi^7_{21}$ [85, (10.6)] is a monomorphism, Corollary 1.3(4) leads to $G_{15}(\mathbb{S}^6) = \pi_{15}(\mathbb{S}^6)$. In particular, we deduce

$$[\iota_6, \mu_6] = 0. \tag{1.131}$$

We have $[\iota_8, \mu_8] = (E\sigma')\mu_{15} \neq 0$ [85, Theorem 12.6] and $[\iota_9, \mu_9] = \eta_9\mu_{10}\sigma_{19} + \sigma_9\eta_{16}\mu_{17} \neq 0$ [85, (12.21), Theorem 12.7].

In view of [74, Theorem] (see also [75, (2.19)] and [85, Theorem 12.22]), we have $\pi^{10}_{28} = \mathbb{Z}_8\{\xi''\} \oplus \mathbb{Z}_2\{\xi'' \pm \lambda''\} \oplus \mathbb{Z}_2\{\eta_{10}\bar{\mu}_{11}\}$. Next, [74, Proposition 1](2), (3) leads to $2\xi'' = x\sigma_{10}\zeta_{17}$ for some odd integer x. Because $\sharp\xi'' = 8$, we derive that $\sharp(\sigma_{10}\zeta_{17}) = 4$. Hence, by [85, (12.25)], we get

$$[\iota_{10}, \mu_{10}] = 2\sigma_{10}\zeta_{17} \neq 0. \tag{1.132}$$

By means of Example 1.19, it holds $[\iota_{11}, \mu_{11}] = 0$. Further, we have $[\iota_{12}, \mu_{12}] \neq 0$ [85, Lemma 16.2] and $[\iota_{13}, \mu_{13}] \neq 0$ [49, p. 309]. By [49, pp. 321–322], we get $[\iota_{14}, \mu_{14}] \neq 0$ and, by [73, p. 140: (8.31), Theorem 3(b)], it holds $[\iota_{22}, \mu_{22}] \neq 0$. Hence, Lemma 1.35(1) and [44, Theorem C] yield

$$\sharp[\iota_n, \mu_n] = \begin{cases} 1, & \text{if } n = 6 \text{ or } n \equiv 3 \pmod 4; \\ 2, & \text{if } n \equiv 0, 1, 2 \pmod 4 \text{ and } n \geq 4 \text{ unless } n = 6. \end{cases} \tag{1.133}$$

We have $[\iota_4, \eta_4\mu_5] = (E\nu')\eta_7\mu_8 \neq 0$ and $[\iota_5, \eta_5\mu_6] = \nu_5\eta_8^2\mu_{10} = 4\nu_5\zeta_8 = 0$ (1.64)(2), [85, Theorem 10.3]. That is,

$$[\iota_5, \eta_5\mu_6] = 0. \tag{1.134}$$

By (1.15) and (1.88), $[\iota_n, \eta_n\mu_{n+1}] = 0$ for $n = 6, 10$ and 11. In view of [85, Theorem 12.7],

$$[\iota_8, \eta_8\mu_9] = (E\sigma')\eta_{15}\mu_{16} \neq 0. \tag{1.135}$$

Further, notice that (1.62), (1.63), (1.64)(2), and (1.70) lead to

$$[\iota_9, \eta_9\mu_{10}] = 4\zeta_9\sigma_{20} + 4\sigma_9\zeta_{16}.$$

Because [85, Lemma 12.12] yields $4\zeta_9\sigma_{20} = 8\sigma_9\zeta_{16} = 0$, in view of [85, Theorem 12.8], we get

$$[\iota_9, \eta_9\mu_{10}] = 4\sigma_9\zeta_{16} \neq 0. \tag{1.136}$$

By Lemma 1.35(2) (see also [49, (7.8)]), we get $[\iota_{12}, \eta_{12}\mu_{13}] \neq 0$ and [49, p. 321] yields $[\iota_{13}, \eta_{13}\mu_{14}] = 8\rho_{13}\sigma_{28} \neq 0$. Next, by [73, p. 139: (8.27)], it holds $[\iota_{21}, \eta_{21}\mu_{22}] \neq 0$.

Hence, by Lemma 1.35(2) and [44, Theorem C],

$$\sharp[\iota_n, \eta_n\mu_{n+1}] = \begin{cases} 1, & \text{if } n = 5 \text{ or } n \equiv 2, 3 \pmod 4; \\ 2, & \text{if } n \equiv 0, 1 \pmod 4 \text{ and } n \geq 4 \text{ unless } n = 5. \end{cases} \tag{1.137}$$

We recall $\pi_{15}(\mathbb{S}^6) = \{\nu_6^3, \mu_6, \eta_6\varepsilon_7\} \cong (\mathbb{Z}_2)^3$. Since $[\iota_6, \eta_6] = 0$ and $\nu_6^3 = \eta_6\bar{\nu}_7$ (1.68), we have $[\iota_6, \nu_6^3] = [\iota_6, \eta_6\varepsilon_7] = 0$. Then, by (1.131), we obtain $G_{15}(\mathbb{S}^6) = \pi_{15}(\mathbb{S}^6)$.

Next, we recall $\pi_{19}(\mathbb{S}^{10}) = \{[\iota_{10}, \iota_{10}], \nu_{10}^3, \mu_{10}, \eta_{10}\varepsilon_{11}\} \cong \mathbb{Z} \oplus (\mathbb{Z}_2)^3$. By (1.88) and (1.68), $[\iota_{10}, \nu_{10}^3] = [\iota_{10}, \eta_{10}\varepsilon_{11}] = 0$. Then, by (1.132), $G_{19}(\mathbb{S}^{10}) = \{3[\iota_{10}, \iota_{10}], \nu_{10}^3, \eta_{10}\varepsilon_{11}\} \cong 3\mathbb{Z} \oplus (\mathbb{Z}_2)^2$.

Let $n \equiv 2 \pmod 4$ and $n \geq 14$. Then, by (1.88),

$$[\iota_n, \eta_n^2\sigma_{n+2}] = [\iota_n, \eta_n\varepsilon_{n+1}] = 0.$$

By (1.133), $[\iota_n, \mu_n] \neq 0$. Whence, we obtain

$$G_{n+9}(\mathbb{S}^n) = \{\nu_n^3, \eta_n\varepsilon_{n+1}\} \cong (\mathbb{Z}_2)^2, \text{ if } n \equiv 2 \pmod 4 \text{ and } n \geq 14.$$

Let $n \equiv 3 \pmod 4$ and $n \geq 11$. Then, by (1.15) and Example 1.19,

$$G_{n+9}(\mathbb{S}^n) = \pi_{n+9}(\mathbb{S}^n), \text{ if } n \equiv 3 \pmod 4.$$

We recall $\pi_{13}(\mathbb{S}^4) = \{\nu_4^3, \mu_4, \eta_4\varepsilon_5\} \cong (\mathbb{Z}_2)^3$. We have $[\iota_4, \nu_4^3] = 2\nu_4^2 \circ \nu_{10}^2 = 0$ and $[\iota_4, \eta_4\varepsilon_5] = (E\nu')\eta_7\varepsilon_8 \neq 0$ [85, Theorem 7.6]. Then, by (1.130), $G_{13}(\mathbb{S}^4) = \{\nu_4^3\} \cong \mathbb{Z}_2$.

Let now $n \equiv 4 \pmod 8$ and $n \geq 12$. By Lemma 1.2(1) and (1.75), we have $[\iota_n, \nu_n^3] = 0$. In light of (1.133) and (1.110), $[\iota_n, \eta_n\varepsilon_{n+1}] = [\iota_n, \eta_n^2\sigma_{n+2}] \neq 0$ and $[\iota_n, \mu_n] \neq 0$. Suppose that $P(\alpha_{2n+1} + \mu_{2n+1}) = 0$ for $\alpha_{2n+1} = \eta_{2n+1}\varepsilon_{2n+2}$ or $\eta_{2n+1}^2\sigma_{2n+3}$. By [85, Proposition 11.10(i)], there exists an element $\beta \in \pi_{2n+7}^{n-1}$ satisfying $E\beta = 0$ and $H\beta = \eta_{2n-3}(\alpha_{2n-2} + \mu_{2n-2}) = \eta_{2n-3}\mu_{2n-2}$. On the other hand, $(\mathcal{P}\mathcal{E}_{2n+7}^{n-1})$ implies a contradictory relation $\beta \in P\pi_{2n+9}^{2n-1} = 0$. So, $[\iota_n, \alpha_n] \neq [\iota_n, \mu_n]$ and hence

$$G_{n+9}(\mathbb{S}^n) = \{\nu_n^3\} \cong \mathbb{Z}_2, \quad \text{if} \quad n \equiv 4 \pmod 8.$$

By (1.45), (1.68), and (1.70) $[\iota_9, \nu_9^3] = (\eta_9^2\sigma_{11} + \sigma_9\eta_{16}^2) \circ \bar{\nu}_{18} = 0$. By (1.44) and (1.83), $[\iota_9, \sigma_9\eta_{16}^2] = \sigma_9(\sigma_{16}\eta_{23}^3) = 4\sigma_9^2\nu_{23} = 0$. So, we obtain $G_{18}(\mathbb{S}^9) = \{\sigma_9\eta_{16}^2, \nu_9^3, \eta_9\varepsilon_{10}\} \cong (\mathbb{Z}_2)^3$. Let now $n \equiv 1 \pmod 8$ and $n \geq 17$. By (1.133), $[\iota_n, \mu_n] \neq 0$ and by (1.34), $[\iota_n, \eta_n\varepsilon_{n+1}] = 0$. In light of IV, $[\iota_n, \nu_n^3] = 0$ if $n = 2^i - 7$ for $i \geq 4$ and $[\iota_n, \nu_n^3] = [\iota_n, \eta_n^2\sigma_{n+2}] \neq 0$ if $n \equiv 1 \pmod 8$ and $n \geq 17$ and $n \neq 2^i - 7$. We show $[\iota_n, \eta_n^2\sigma_{n+2}] \neq [\iota_n, \mu_n]$. Suppose otherwise. Then, by [85, Proposition 11.10(ii)], there is an element $\beta \in \pi_{2n+7}^{n-1}$ such that $E\beta = P(\eta_{2n+1}^2\sigma_{2n+2} + \mu_{2n+1}) = 0$ and $H\beta = \eta_{2n-3}\mu_{2n-2}$. On the other hand, by $(\mathcal{PE}_{2n+7}^{n-1})$ and (1.76), $H\beta = 0$, and so we get the assertion. Hence, we obtain

$$G_{n+9}(\mathbb{S}^n) = \begin{cases} \{\eta_n\varepsilon_{n+1}\} \cong \mathbb{Z}_2, & \text{if } n \equiv 1 \pmod 8 \text{ and } n \geq 17 \text{ and } n \neq 2^i - 7; \\ \{\eta_n\varepsilon_{n+1}, \nu_n^3\} \cong (\mathbb{Z}_2)^2, & \text{if } n = 2^i - 7 \ (i \geq 5). \end{cases}$$

By (1.33) and [85, (7.10)], $[\iota_5, \eta_5\varepsilon_6] = \nu_5\eta_8^2\varepsilon_{10} = 4\nu_5^2\sigma_{11} = 0$. So, we obtain $G_{14}(\mathbb{S}^5) = \{\nu_5^3, \eta_5\varepsilon_6\} \cong (\mathbb{Z}_2)^2$. Let $n \equiv 5 \pmod 8$ and $n \geq 13$. By Proposition 1.36 and (1.133), $\nu_n^3 \in G_{n+9}(\mathbb{S}^n)$ and $\mu_n \notin G_{n+9}(\mathbb{S}^n)$. Furthermore, by Proposition 1.36, $\eta_n\varepsilon_{n+1} \in G_{n+9}(\mathbb{S}^n)$ unless $n \equiv 53 \pmod{64}$. So, we obtain

$$G_{n+9}(\mathbb{S}^n) = \{\nu_n^3, \eta_n\varepsilon_{n+1}\} \cong (\mathbb{Z}_2)^2, \text{ if } n \equiv 5 \pmod 8 \text{ and } n \not\equiv 53 \pmod{64}.$$

Now, recall that in view of [85, Theorem 7.4], the map $E^\infty : \pi_{22}^{11} \to \pi_{11}^s$ is an isomorphism. Then, the relations [85, (9.3)] and (1.84) yield

$$x\zeta_n \in \{\nu_n, \sigma_{n+3}, 16\iota_{n+10}\} \text{ for some odd integer } x \text{ and } n \geq 11. \quad (1.138)$$

Further, [85, Lemma 9.1] leads to

$$\zeta_n \in \{2\iota_n, \eta_n, \alpha_{n+1}\}_2 \pmod{2\zeta_n} \text{ for } \alpha_{n+1} = \eta_{n+1}^2\sigma_{n+3} \text{ or } \eta_{n+1}\varepsilon_{n+2}, \text{ if } n \geq 11.$$

Let $n \equiv 0 \pmod 8$ and $n \geq 16$. By [85, Proposition 11.11(i)], there exists an element $\beta \in \pi_{2n+6}^{n-2}$ such that $[\iota_n, \alpha_n] = E^2\beta$ and $H\beta \in \{2\iota_{2n-5}, \eta_{2n-5}, \alpha_{2n-4}\}_2 \ni \zeta_{2n-5} \pmod{2\zeta_{2n-5}}$. Suppose that $[\iota_n, \alpha_n] = 0$. Then, $(\mathcal{PE}_{2n+7}^{n-1})$ induces a relation $E\beta \in P\pi_{2n+9}^{2n-1} = 0$. By $(\mathcal{PE}_{2n+6}^{n-2})$ and (1.76), we have a contradictory relation $\zeta_{2n-5} \in 2\pi_{2n+6}^{2n-5}$. Whence, we get that $[\iota_n, \alpha_n] \neq 0$. In light of (1.133) and (1.137), we know $[\iota_n, \mu_n] \neq 0$ and $[\iota_n, \mu_n]\eta_{2n+8} \neq 0$. This implies that $[\iota_n, \alpha_n] \neq [\iota_n, \mu_n]$ and $[\iota_n, \nu_n^3] \neq [\iota_n, \mu_n]$.

By (1.22) and (1.107)(2), $[\iota_8, \nu_8^3] = (E\sigma')\nu_{15}^3 = \eta_8\bar{\varepsilon}_9$ and $[\iota_8, \sigma_8\eta_{16}^2] = (E\sigma')\sigma_{15}\eta_{22}^2 = (E\sigma')(\eta_{15}\varepsilon_{16} + \nu_{15}^3) = [\iota_8, \eta_8\varepsilon_9] + [\iota_8, \nu_8^3]$. We have $[\iota_8, (E\sigma')\eta_{15}^2] = 0$. So, we obtain $G_{17}(\mathbb{S}^8) = \{(E\sigma')\eta_{15}^2, \sigma_8\eta_{15}^2 + \nu_8^3 + \eta_8\varepsilon_9\} \cong (\mathbb{Z}_2)^2$. By [73, p. 71], $\mathrm{Ker}\{P : \pi_{42}^{33} \to \pi_{40}^{16}\} = 0$ and hence $G_{25}(\mathbb{S}^{16}) = 0$.

Let $n \equiv 2 \pmod 4$ and $n \geq 6$. By (1.64)(2), Lemma 1.2(1) and (1.16), $4[\iota_n, \zeta_n] = 0$. Then, by the relation $H[\iota_n, \zeta_n] = \pm 2\zeta_{2n-1}$, we obtain

$$\sharp[\iota_n, \zeta_n] = 4, \text{ if } n \equiv 2 \pmod 4 \text{ and } n \geq 6. \quad (1.139)$$

By [69, 4.14], there exists an element $\tau_1 \in \pi^{n-6}_{2n+2}$ such that

$$[\iota_n, \nu_n^3] = E^6\tau_1,\ H\tau_1 = \eta_{2n-13}\kappa_{2n-12},\ \text{if } n \equiv 0 \pmod 8 \text{ and } n \geq 16.$$

Suppose that $[\iota_n, \nu_n^3] = 0$. Then, by $(\mathcal{PE}^{n-1}_{2n+7})$, we have $E^5\tau_1 = 0$. Then, by $(\mathcal{PE}^{n-2}_{2n+6})$, we have $E^4\tau_1 \in P\pi^{2n-3}_{2n+8} = \{[\iota_{n-2}, \zeta_{n-2}]\}$. By applying $H\colon \pi^{n-2}_{2n+6} \to \pi^{2n-5}_{2n+6}$ to this relation and by (1.139), we obtain $E^4\tau_1 = 4a[\iota_{n-2}, \zeta_{n-2}] = 0$ for $a \in \{0, 1\}$. By the fact that $\pi^{2n-5}_{2n+7} = \pi^{2n-7}_{2n+6} = 0$, we obtain $E^2\tau_1 = 0$. Hence, by $(\mathcal{PE}^{n-5}_{2n+3})$ and (1.91), we have

$$E\tau_1 \in P\pi^{2n-9}_{2n+5} = E^3\bar{\tau}_{n-8} \circ \{\sigma^2_{2n-11}, \kappa_{2n-11}\}.$$

By $(\mathcal{PE}^{n-6}_{2n+2})$, we obtain

$$\tau_1 + E^2(b\bar{\tau}_{n-8}\sigma^2_{2n-14} + b\bar{\tau}_{n-8}\kappa_{2n-14}) \in P\pi^{2n-11}_{2n+4} \text{ with } b, c \in \{0, 1\}.$$

This induces a contradictory relation $\eta_{2n-13}\kappa_{2n-12} \in 2\pi^{2n-13}_{2n+2}$. Thus, we conclude that

$$[\iota_n, \nu_n^3] \neq 0,\ \text{if } n \equiv 0 \pmod 8 \text{ and } n \geq 16.$$

Summing the above, we get:

Proposition 1.46. *The group $\mathrm{G}_{n+9}(\mathbb{S}^n)$ is equal to the following: $\pi_{n+9}(\mathbb{S}^n)$ if $n = 6$ or $n \equiv 3 \pmod 4$; $\{\nu_n^3, \eta_n\varepsilon_{n+1}\} \cong (\mathbb{Z}_2)^2$ if $n \equiv 2 \pmod 4$ and $n \geq 14$, $n = 2^i - 7$ for $i \geq 5$ or $n \equiv 5 \pmod 8$ unless $n \equiv 53 \pmod{64}$; $\{\nu_n^3\} \cong \mathbb{Z}_2$ if $n \equiv 4 \pmod 8$ or $53 \pmod{64}$ and $n \geq 117$; $\{\eta_n\varepsilon_{n+1}\} \cong \mathbb{Z}_2$ if $n \equiv 1 \pmod 8$ and $n \geq 17$ and $n \neq 2^i - 7$; 0 if $n \equiv 0 \pmod 8$ and $n \geq 16$. Moreover, $\mathrm{G}_{17}(\mathbb{S}^8) = \{(E\sigma')\eta^2_{15}, \sigma_8\eta^2_{15} + \nu_8^3 + \eta_8\varepsilon_9\} \cong (\mathbb{Z}_2)^2$, $\mathrm{G}_{18}(\mathbb{S}^9) = \{\sigma_9\eta^2_{16}, \nu_9^3, \eta_9\varepsilon_{10}\} \cong (\mathbb{Z}_2)^3$ and $\mathrm{G}_{19}(\mathbb{S}^{10}) = \{3[\iota_{10}, \iota_{10}], \nu_{10}^3, \eta_{10}\varepsilon_{11}\} \cong 3\mathbb{Z} \oplus (\mathbb{Z}_2)^2$.*

Now, by (1.1), Corollary 1.3(3), Propositions 1.8, (1.7) and (1.137), we can determine $G_{n+10}(\mathbb{S}^n)$ for $n \geq 12$.

We have $G_{14}(\mathbb{S}^4; 5) = \pi_{14}(\mathbb{S}^4; 5) \cong \mathbb{Z}_5$ and $G_{14}(\mathbb{S}^4; 3) = \pi_{14}(\mathbb{S}^4; 3) \cong (\mathbb{Z}_3)^2$ by (1.8).

By [85, Theorem 7.3], $\pi^4_{14} = \mathbb{Z}_8\{\nu_4\sigma'\} \oplus \mathbb{Z}_4\{E\varepsilon'\} \oplus \mathbb{Z}_2\{\eta_4\mu_5\}$. We have $[\iota_4, \nu_4\sigma'] = 2\nu_4^2E^3\sigma'$ and $[\iota_4, E\varepsilon'] = 2\nu_4E^4\varepsilon' - E(\nu'E^3\varepsilon') = 2\nu_4E^4\varepsilon'$ (1.124). In view of [85, (7.10), (7.16)], it holds $\nu_5E\sigma' = 2(\nu_5\sigma_8) = \pm E^2\varepsilon'$. Therefore, we conclude that $\nu_4\sigma' \pm E\varepsilon' \in G_{14}(\mathbb{S}^4)$. We also obtain $2E\varepsilon' \in G_{14}(\mathbb{S}^4)$, because $[\iota_4, 2E\varepsilon'] = 4(\nu_4E^4\varepsilon') = 0$. Then, (1.24) and (1.134) imply $G_{15}(\mathbb{S}^5) = \pi_{15}(\mathbb{S}^5)$.

We recall the following:

$$\pi_{16}(\mathbb{S}^6) = \{\nu_6\sigma_9, \eta_6\mu_7, \beta_1(6)\} \cong \mathbb{Z}_{72} \oplus \mathbb{Z}_2,$$

$$\pi_{18}(\mathbb{S}^8) = \{\sigma_8\nu_{15}, \nu_8\sigma_{11}, \eta_8\mu_9, \sigma_8\alpha_1(15), \beta_1(8)\} \cong (\mathbb{Z}_{24})^2 \oplus \mathbb{Z}_2,$$

$$\pi^9_{19} = \{\sigma_9\nu_{16}, \eta_9\mu_{10}\} \cong \mathbb{Z}_8 \oplus \mathbb{Z}_2,$$

$$\pi^{10}_{20} = \{\sigma_{10}\nu_{17}, \eta_{10}\mu_{11}\} \cong \mathbb{Z}_4 \oplus \mathbb{Z}_2,\ \pi^{11}_{21} = \{\sigma_{11}\nu_{18}, \eta_{11}\mu_{12}\} \cong (\mathbb{Z}_2)^2.$$

The order $\sharp[\iota_6, \beta_1(6)] = \sharp[\iota_6, \iota_6] \circ \beta_1(11) = 3$. By (1.15), $[\iota_6, \eta_6\mu_7] = 0$. By (1.83), $[\iota_6, \nu_6\sigma_9] = [\iota_6, \iota_6](\nu_{11}\sigma_{14}) = 0$. This yields $G_{16}(\mathbb{S}^6) = 3\pi_{16}(\mathbb{S}^6)$.

It holds that $[\iota_8, \beta_1(8)] \neq 0$ and $[\iota_8, \sigma_8\alpha_1(15)] = [\iota_8, \iota_8](\alpha_2(15)\alpha_1(22)) = 0$ (1.8). By (1.83), $[\iota_8, \sigma_8\nu_{15}] = [\iota_8, \nu_8\sigma_{11}] = 0$. Hence, by (1.135), we get that $G_{18}(\mathbb{S}^8) = \{\sigma_8\nu_{15}, \nu_8\sigma_{11}, \sigma_8\alpha_1(15)\} \cong (\mathbb{Z}_8)^2 \oplus \mathbb{Z}_3$.

We have $[\iota_9, \sigma_9\nu_{16}] = 0$. Then, by (1.136) and Corollary 1.3(3), $G_{19}(\mathbb{S}^9) = \{\sigma_9\nu_{16}, \beta_1(9)\} \cong \mathbb{Z}_{24}$.

We obtain $[\iota_{10}, \sigma_{10}\nu_{17}] = 0$ by (1.83), $[\iota_{10}, \eta_{10}\mu_{11}] = 0$ by (1.88) and hence $G_{20}(\mathbb{S}^{10}) = \pi^{10}_{20}$.

By (1.15) and (1.46), $[\iota_{11}, \eta_{11}\mu_{12}] = [\iota_{11}, \sigma_{11}\nu_{18}] = 0$. This yields $G_{21}(\mathbb{S}^{11}) = \pi_{21}(\mathbb{S}^{11})$.

Therefore, we conclude that

$$G_{n+10}(\mathbb{S}^n) = \begin{cases} \{\nu_4^+\sigma'^+ \pm E\varepsilon', 2E\varepsilon', \alpha_1(4)\alpha_2(7), & \\ \nu_4\alpha_2(7), \nu_4\alpha'_1(7)\}, & \text{if } n = 4; \\ \pi_{15}(\mathbb{S}^5), & \text{if } n = 5; \\ \pi^6_{16} \oplus \{3\beta_1(6)\}, & \text{if } n = 6; \\ \{\sigma_8\nu_{15}, \nu_8\sigma_{11}, \sigma_8^+\alpha_1(15)\}, & \text{if } n = 8; \\ \{\sigma_9\nu_{16}, \beta_1(9)\}, & \text{if } n = 9; \\ \pi^{10}_{20} = \{\sigma_{10}\nu_{17}, \eta_{10}\mu_{11}\}, & \text{if } n = 10; \\ \pi_{21}(\mathbb{S}^{11}), & \text{if } n = 11. \end{cases}$$

Thus, by summing up the above results, we get

Proposition 1.47. *The group* $G_{n+10}(\mathbb{S}^n)$ *is isomorphic to the following:* $\mathbb{Z}_{120} \oplus \mathbb{Z}_6$, $\mathbb{Z}_{72} \oplus \mathbb{Z}_2$, $\mathbb{Z}_{24} \oplus \mathbb{Z}_2$, $\mathbb{Z}_{24} \oplus \mathbb{Z}_8$, $\mathbb{Z}_{24}$, $\mathbb{Z}_4 \oplus \mathbb{Z}_2$, $\mathbb{Z}_6 \oplus \mathbb{Z}_2$ *according as* $n = 4, 5, 6, 8, 9, 10, 11$. *Furthermore,* $G_{n+10}(\mathbb{S}^n)$ *is isomorphic to the group:* 0 *if* $n \equiv 0 \pmod 4$ *and* $n \geq 12$; $\mathbb{Z}_2$ *if* $n \equiv 2 \pmod 4$ *and* $n \geq 14$; $\mathbb{Z}_3$ *if* $n \equiv 1 \pmod 4$ *and* $n \geq 13$, *and* $\mathbb{Z}_6$ *if* $n \equiv 3 \pmod 4$ *and* $n \geq 15$.

By [85, (10.14)], $[\iota_5, \zeta_5] = 0$. By (1.139), $\sharp[\iota_6, \zeta_6] = \sharp[\iota_{10}, \zeta_{10}] = 4$. In view of [85, Theorem 12.8, Lemma 12.12], $\sharp[\iota_8, \zeta_8] = 8$. By means of [85, (12.22)], $E\colon \pi^9_{28} \to \pi^{10}_{29}$ is an isomorphism, and so $[\iota_9, \zeta_9] = 0$. By [49, pp. 307, 320], $[\iota_{11}, \zeta_{11}] = 0$ and $\sharp[\iota_{12}, \zeta_{12}] = 8$. Further, by [52, (3.10)], $[\iota_{13}, \zeta_{13}] = 0$. Summing up these results, we get that $\sharp[\iota_n, \zeta_n] = 1, 4, 8, 1, 4, 1, 8, 1$ according as $n = 5, 6, 8, 9, 10, 11, 12, 13$.

By (1.64)(2), we have $[\iota_4, E\mu'] = 4\nu_4\zeta_7 \neq 0$ and (1.123) leads to $[\iota_4, \varepsilon_4\nu_{12}] = [\iota_4, \iota_4](\varepsilon_7\nu_{15}) = 0$. We note that $[\iota_6, \bar{\nu}_6] = 0$ (1.117) and $[\iota_n, \bar{\nu}_n\nu_{n+8}] = 0$ for $n = 8, 9$ by (1.21). Hence, by the group structure of π^n_{n+11} [85, Theorem 7.4], we obtain $G_{n+11}(\mathbb{S}^n; 2)$ for $5 \leq n \leq 12$. Summing up, we obtain

$$G_{n+11}(\mathbb{S}^n) = \begin{cases} \{\nu_4\sigma'\eta_{14}, \nu_4\bar{\nu}_7, \nu_4\varepsilon_7, \\ 2E\mu', \varepsilon_4\nu_{12}, (E\nu')\varepsilon_7\}, & \text{if } n = 4; \\ \pi_{16}(\mathbb{S}^5), & \text{if } n = 5; \\ \{4\zeta_6, \bar{\nu}_6\nu_{14}\}, & \text{if } n = 6; \\ \{\bar{\nu}_8\nu_{16}\}, & \text{if } n = 8; \\ \pi_{20}(\mathbb{S}^9), & \text{if } n = 9; \\ 4\pi^{10}_{21}, & \text{if } n = 10; \\ \pi_{22}(\mathbb{S}^{11}), & \text{if } n = 11; \\ \{3[\iota_{12}, \iota_{12}]\}, & \text{if } n = 12. \end{cases}$$

By abuse of notations, ζ_n for $n \ge 5$ represents a generator of the direct summands $\mathbb{Z}_8$ of π^n_{n+11} and $\mathbb{Z}_{504}$ of $\pi_{n+11}(\mathbb{S}^n)$, respectively.

Notice that the result $\eta_4\zeta_5 \equiv (E\nu')\mu_7 \pmod{(E\nu')\eta_7\varepsilon_8}$ [75, Proposition (2.2)(6)] yields

$$\eta_5\zeta_6 = 0. \tag{1.140}$$

We already know $[\iota_5, \zeta_5] = 0$ and $\sharp[\iota_{12}, \zeta_{12}] = 8$. By [73, p. 139: (8.24)], $\sharp[\iota_{20}, \zeta_{20}] = 8$. Hence, by [44, Theorem C], Corollary 1.3(3), (1.7), Theorem 1.43, and (1.139), we obtain

$$\sharp[\iota_n, \zeta_n] = \begin{cases} 1, & \text{if } n \equiv 1 \pmod 2 \text{ and } n \ge 5 \text{ unless } n \equiv 115 \pmod{128}; \\ 2, & \text{if } n \equiv 115 \pmod{128} \text{ and } n \ge 243; \\ 252, & \text{if } n \equiv 2 \pmod 4 \text{ and } n \ge 6; \\ 504, & \text{if } n \equiv 0 \pmod 4 \text{ and } n \ge 8. \end{cases}$$

Thus, by summing up the above results, we get

Proposition 1.48. *The group $G_{n+11}(\mathbb{S}^n)$ is isomorphic to the following: $(\mathbb{Z}_2)^6$, $\mathbb{Z}_{504} \oplus (\mathbb{Z}_2)^2$, $\mathbb{Z}_2 \oplus \mathbb{Z}_4$, $\mathbb{Z}_2$, $\mathbb{Z}_{504} \oplus \mathbb{Z}_2$, $\mathbb{Z}_2$, $\mathbb{Z}_{504}$, $3\mathbb{Z}$ according as $n = 4, 5, 6, 8, 9, 10, 11, 12$. Furthermore, $G_{n+11}(\mathbb{S}^n)$ is isomorphic to the group: $\mathbb{Z}_{504}$ if $n \equiv 1 \pmod 2$ and $n \ge 13$ unless $n \equiv 115 \pmod{128}$; $\mathbb{Z}_{252}$ if $n \equiv 115 \pmod{128}$ and $n \ge 243$; $\mathbb{Z}_2$ if $n \equiv 2 \pmod 4$ and $n \ge 14$ and 0 if $n \equiv 0 \pmod 4$ and $n \ge 16$.*

By the use of [85, Theorem 7.6, p. 187: Table], we obtain $G_{n+12}(\mathbb{S}^n) = \pi_{n+12}(\mathbb{S}^n)$ for $n \le 9$.

We recall $\pi_{22}(\mathbb{S}^{10}) = \{[\iota_{10}, \nu_{10}]\} \cong \mathbb{Z}_{12}$. By Proposition 1.10(1), $G_{22}(\mathbb{S}^{10}; 3) = 0$ and hence $G_{22}(\mathbb{S}^{10}) = \pi^{10}_{22}$. By [49, (7.7)], $G_{23}(\mathbb{S}^{11}) = \pi_{23}(\mathbb{S}^{11})$. By [85, (7.30)] and [52, (4.29)], we obtain $G_{n+12}(\mathbb{S}^n) = \pi_{n+12}(\mathbb{S}^n)$ for $n = 12$ and 13. Summing up, we obtain

$$G_{n+12}(\mathbb{S}^n) = \pi_{n+12}(\mathbb{S}^n) \text{ unless } n = 10 \text{ and } G_{22}(\mathbb{S}^{10}) = \pi^{10}_{22}.$$

By the use of [85, Theorem 7.7, pp. 187–188: Table], we obtain $G_{n+13}(\mathbb{S}^n)$. In particular, we need the relations: $[\iota_{11}, \theta'] = 0$ and $[\iota_{12}, \theta] = 0$ for $\theta' \in \pi_{23}^{11}$ and $\theta \in \pi_{24}^{12}$. We show the case $n = 4$. We recall

$$\pi_{17}(\mathbb{S}^4) = \{\nu_4^2\sigma_{10}, \nu_4\eta_7\mu_8, (E\nu')\eta_7\mu_8, \nu_4\beta_1(7), \alpha_1(4)\beta_1(7)\} \cong \mathbb{Z}_{24} \oplus \mathbb{Z}_6 \oplus \mathbb{Z}_2.$$

We have $G_{17}(\mathbb{S}^4; 2) = \pi_{17}^4$. We see that $[\iota_4, \nu_4\beta_1(7)] = \pm 2\nu_4\alpha_1(7)\beta_1(10)$ and $[\iota_4, \alpha_1(4)\beta_1(7)] = \pm(2\nu_4 + \alpha_1(4))(\alpha_1(7)\beta_1(10))$. By making use of the exact sequence in [85, Proposition 13.3], we have $\pi_{19}(\mathbb{S}^3; 3) = \{\alpha_1(3)\alpha_1(6)\beta_1(9)\} \cong \mathbb{Z}_3$. So, $[\iota_4, \nu_4\beta_1(7)]$ and $[\iota_4, \alpha_1(4)\beta_1(7)]$ generate the group $\pi_{20}(\mathbb{S}^4; 3) \cong (\mathbb{Z}_3)^2$ and hence $G_{17}(\mathbb{S}^4; 3) = 0$.

Summing up, we obtain

$$G_{n+13}(\mathbb{S}^n) = \begin{cases} \pi_{n+13}(\mathbb{S}^n), & \text{if } n \text{ is odd or } n = 2; \\ \pi_{n+13}^n, & \text{if } n \text{ is even unless } n = 2, 14; \\ \{3[\iota_{14}, \iota_{14}]\} \cong 3\mathbb{Z}, & \text{if } n = 14. \end{cases}$$

Chapter 2
Gottlieb and Whitehead Center Groups of Projective Spaces

By the use of Siegel's method and the classical results of homotopy groups of spheres and Lie groups, we determine in this chapter some Gottlieb groups of projective spaces or give the lower bounds of their orders. Furthermore, making use of the properties of Whitehead products, we determine some Whitehead center groups of projective spaces.

2.1 Preliminaries I

Let EX be the suspension of a space X and denote by $E : \pi_k(X) \to \pi_{k+1}(EX)$ the suspension homomorphism. Next, write $\sharp\alpha$ for the order of $\alpha \in \pi_k(X)$ and $\Delta = \Delta_{\mathbb{F}P} : \pi_k(\mathbb{F}P^n) \to \pi_{k-1}(\mathbb{S}^{d-1})$ for the connecting map. By [8, Theorem (2.1)] it holds:

$$\Delta(i_{\mathbb{F}*}E) = \mathrm{id}_{\pi_{k-1}(\mathbb{S}^{d-1})} \text{ and } \pi_k(\mathbb{F}P^n) = \gamma_{n*}\pi_k(\mathbb{S}^{d(n+1)-1}) \oplus i_{\mathbb{F}*}E\pi_{k-1}(\mathbb{S}^{d-1}).$$

Especially,

$$\sharp\alpha = \sharp(i_{\mathbb{H}}E\alpha) \text{ for } \alpha \in \pi_{k-1}(\mathbb{S}^3). \tag{2.1}$$

According to [88], a map $f : X \to Y$ is *cyclic* if $f \vee \mathrm{id}_Y : X \vee Y \to Y$ extends to $F : X \times Y \to Y$. The extension F is called an *associated map* for f [39]. We recall from [81, Lemma 2.1] and [88, Lemma 1.3]:

Lemma 2.1 (Siegel-Varadarajan). *Let $f : X \to Y$ be cyclic. Then, the composite $f \circ g : W \to Y$ is cyclic for any map $g : W \to X$.*

In particular, elements of $G_n(X)$ are cyclic. Notice that, in view of Chap. 1, an element $\alpha \in G_n(\mathbb{S}^m)$ if and only if $[\iota_m, \alpha] = 0$.

© Springer International Publishing Switzerland 2014

M. Golasiński, J. Mukai, *Gottlieb and Whitehead Center Groups of Spheres, Projective and Moore Spaces*, DOI 10.1007/978-3-319-11517-7_2

We say that a subgroup $A_n(X) \subseteq \pi_n(X)$ is *admissible* if $A_n(X) \circ \pi_k(\mathbb{S}^n) \subseteq A_k(X)$ for all $k \geq 1$. Further, an element $\alpha \in A_n(X)$ is called $A_n(X)$*-admissible.*

Recall that given $\alpha \in \pi_k(\mathbb{S}^m)$ and $\beta \in \pi_l(\mathbb{S}^n)$ with $m, n > 0$, Lemma 1.2(1) yields

$$[\xi \circ \alpha, \eta \circ \beta] = 0 \text{ provided that } [\xi, \eta] = 0 \tag{2.2}$$

for $\xi \in \pi_m(X)$, $\eta \in \pi_n(X)$.

Then, (2.2) and Lemma 2.1 yield:

Corollary 2.2. *The subgroups* $G_n(X), P_n(X) \subseteq \pi_n(X)$ *are admissible for all* $n \geq 1$.

To state an analog of the formula from Lemma 1.2(2) for $m > n = 0$, we recall [92, Chap. III, Sect. 1] that given a space (X, x_0) with a base point and a path $u : I \to Y$ in a space Y, maps $f, g : X \to Y$ are called to be *freely homotopic along* u ($f \simeq_u g$) if there is such a homotopy $H : I \times X \to Y$ that $H(t, x_0) = u(t)$ for all $t \in I$.

Lemma 2.3. *If* $f \simeq_u g$ *then* $f \circ f' \simeq_u g \circ f'$ *for any map* $f' : (X', x'_0) \to (X, x_0)$ *of spaces with base points.*

Proof. Given a homotopy $H : I \times X \to Y$ with $H(t, x_0) = u(t)$ for all $t \in I$, consider $H' : I \times X' \to Y$ given by $H'(t, x') = H(t, f'(x'))$ for $(t, x') \in I \times X'$. Then, $H'(t, x'_0) = u(t)$ for all $t \in I$ and the result follows. □

Now, let (X, x_0) be a space with *nondegenerate base point* (i.e., $(X, \{x_0\})$ is an ANR-pair [92, p. 22]), $f : (X, x_0) \to (Y, y_0)$, and $u : I \to Y$ such a path in Y that $f(x_0) = u(0)$. Then, by [92, Chap. III, (1.4)], there is such $g : (X, x_0) \to (Y, u(1))$ that $f \simeq_u g$.

Next, write η and ξ for the homotopy classes of u and f, respectively. Then, in view of [92, Chap. III, (1.5) and (1.6)], the homotopy class $\tau_\eta(\xi)$ of g is well defined. Consequently, Lemma 2.3 leads to

$$\tau_\eta(\xi \circ \xi') = \tau_\eta(\xi) \circ \xi' \tag{2.3}$$

for the homotopy class ξ' of a map $f' : (X', x'_0) \to (X, x_0)$.

On the other hand, for $\eta \in \pi_1(X)$ and $\xi \in \pi_{m+1}(X)$, the element $\tau_\eta(\xi) \in \pi_{m+1}(X)$ is the operation of η on ξ and, in view of [92, Chap. X, (7.6)], it holds

$$[\xi, \eta] = (-1)^{m+1}(\tau_\eta(\xi) - \xi). \tag{2.4}$$

Then, in virtue of (2.3), (2.4), and [92, Chap. X, (8.2) Theorem], we may state

$$[\xi \circ E\alpha, \eta] = (-1)^{k+m}[\xi, \eta] \circ E\alpha \tag{2.5}$$

for $\alpha \in \pi_k(\mathbb{S}^m)$, $\eta \in \pi_1(X)$, and $\xi \in \pi_{m+1}(X)$.

Let $q_n = q_{n,\mathbb{F}} : \mathbb{F}P^n \to \mathbb{S}^{dn}$ be the map pinching $\mathbb{F}P^{n-1}$ to the base point. Then, in view of [8] and [33, (2.10)(a)],

$$q_n\gamma_n = \begin{cases} (1-(-1)^n)\iota_n & \text{for } \mathbb{F}=\mathbb{R}, \\ n\eta_{2n} & \text{for } \mathbb{F}=\mathbb{C}, \\ \pm n\nu_{4n}^{+} & \text{for } \mathbb{F}=\mathbb{H}. \end{cases} \tag{2.6}$$

By (2.2), Proposition 1.6 and [8, (4.1-3)], we obtain a key formula determining the Whitehead center groups of $\mathbb{F}P^n$.

Lemma 2.4. *Let $h_0\alpha \in \pi_k(\mathbb{S}^{2d(n+1)-3})$ be the 0-th Hopf–Hilton invariant for $\alpha \in \pi_k(\mathbb{S}^{d(n+1)-1})$. Then:*

1. $[\gamma_n\alpha, i_{\mathbb{R}}] = \begin{cases} 0 & \text{for odd } n, \\ (-1)^k\gamma_n(-2\alpha + [\iota_n,\iota_n]\circ h_0\alpha) & \text{for even } n; \end{cases}$
2. $[\gamma_n\alpha, i_{\mathbb{C}}] = \begin{cases} 0 & \text{for odd } n, \\ \gamma_n(\eta_{2n+1}\circ E\alpha + [\iota_{2n+1},\eta_{2n+1}]\circ Eh_0\alpha) & \text{for even } n; \end{cases}$
3. $[\gamma_n\alpha, i_{\mathbb{H}}] = \pm(n+1)\gamma_n(\nu_{4n+3}^{+}\circ E^3\alpha + [\iota_{4n+3},\nu_{4n+3}]\circ E^3h_0\alpha)$.

We recall:

Lemma 2.5 (Mimura [49, Lemma 5.9]). *If $\alpha \in \pi_p(\mathbb{S}^n)$ and $\beta \in \pi_i(\mathbb{S}^p)$ then*

$$Eh_0(\alpha\beta) = (Eh_0(\alpha))(E\beta) + (E^n\alpha)(E^p\alpha)(Eh_0(\beta)).$$

Observe that [49, Corollary 5.8, p. 183] implies

$$H(\alpha) = h_0(\alpha) \tag{2.7}$$

for $\alpha \in \pi_k(\mathbb{S}^n)$ provided that $k \le 3n-3$.

Given a fibration $F \hookrightarrow E \xrightarrow{p} B$, write $\partial : \pi_{n+1}(B) \longrightarrow \pi_n(F)$ for the connecting map. If $\partial(\alpha) = 0$ then there exists γ with $\alpha = p_*\gamma$ and $\partial(\alpha\circ\beta) = \partial p_*(\gamma\circ\beta) = 0$. Thus, we get that the subgroup

$$\operatorname{Ker}\{\partial : \pi_{k+1}(B) \longrightarrow \pi_k(F)\} \subseteq \pi_{n+1}(B) \text{ is admissible} \tag{2.8}$$

and its elements are called *∂-cyclic*.

We also recall the formula

$$\partial(\alpha\circ E\beta) = \partial(\alpha)\circ\beta \quad [37, \text{Lemma } 1]. \tag{2.9}$$

We set

$$SO_{\mathbb{F}}(n) = \begin{cases} SO(n) & \text{for } \mathbb{F}=\mathbb{R}; \\ SU(n) & \text{for } \mathbb{F}=\mathbb{C}; \\ Sp(n) & \text{for } \mathbb{F}=\mathbb{H}. \end{cases}$$

Denote by $i_{n,\mathbb{F}} : SO_{\mathbb{F}}(n-1) \hookrightarrow SO_{\mathbb{F}}(n)$ and $p_{n,\mathbb{F}} : SO_{\mathbb{F}}(n) \longrightarrow \mathbb{S}^{dn-1}$ the inclusion and projection maps, respectively. Then, we consider the exact sequence induced by the fibration $SO_{\mathbb{F}}(n+1) \overset{SO_{\mathbb{F}}(n)}{\longrightarrow} \mathbb{S}^{d(n+1)-1}$:

$$(\mathcal{SF}^n_k) \quad \cdots \to \pi_{k+1}(\mathbb{S}^{d(n+1)-1}) \xrightarrow{\Delta_{\mathbb{F}}} \pi_k(SO_{\mathbb{F}}(n)) \xrightarrow{i_*} \pi_k(SO_{\mathbb{F}}(n+1)) \xrightarrow{p_*} \pi_k(\mathbb{S}^{d(n+1)-1}) \to \cdots,$$

where $\Delta_{\mathbb{F}}$ is the connecting map, $i = i_{n+1,\mathbb{F}}$ and $p = p_{n+1,\mathbb{F}}$, respectively. A $\Delta_{\mathbb{F}}$-cyclic element $\alpha \in \pi_k(\mathbb{S}^{d(n+1)-1})$ is called simply $\mathbb{F}$-*cyclic*.

Notice that $\pi_3(SO(4)) = \{[\eta_2]_4, [\iota_3]\} \cong \mathbb{Z}^2$ and any element of $\pi_k(SO(4)) \cong \pi_k(SO(3)) \oplus \pi_k(\mathbb{S}^3)$ is uniquely represented by two elements of $\pi_k(\mathbb{S}^3)$:

$$[\eta_2]_4\alpha + [\iota_3]\beta \quad \text{for} \quad \alpha, \beta \in \pi_k(\mathbb{S}^3),$$

where $[\eta_2]$ and $[\iota_3]$ are lifts of η_2 and ι_3, respectively, defined by (1.56). Certainly, we can take $J([\iota_3]) = \nu_4^+$ and $J([\eta_2]) = \nu'^+$.

Next, define the subgroups of $\pi_k(\mathbb{S}^{d(n+1)-1})$ and $\pi_{k-1}(\mathbb{S}^{d-1})$ as follows:

$$M_k(\mathbb{S}^{d(n+1)-1}) = M_{k,\mathbb{F}}(\mathbb{S}^{d(n+1)-1}) = \{\alpha \in \pi_k(\mathbb{S}^{d(n+1)-1}) \mid [\gamma_n\alpha, i_{\mathbb{F}}] = 0\},$$

$$L'_{k-1}(\mathbb{S}^{d-1}) = L'_{k-1,n}(\mathbb{S}^{d-1}) = \{\beta \in \pi_{k-1}(\mathbb{S}^{d-1}) \mid [i_{\mathbb{F}}E\beta, i_{\mathbb{F}}] = 0\},$$

$$L''_{k-1}(\mathbb{S}^{d-1}) = L''_{k-1,n}(\mathbb{S}^{d-1}) = \{\beta \in \pi_{k-1}(\mathbb{S}^{d-1}) \mid [i_{\mathbb{F}}E\beta, \gamma_n] = 0\},$$

and

$$L_{k-1}(\mathbb{S}^{d-1}) = L_{k-1,n}(\mathbb{S}^{d-1}) = L'_{k-1,n}(\mathbb{S}^{d-1}) \cap L''_{k-1,n}(\mathbb{S}^{d-1}).$$

We also define the subgroup $\mathcal{Q}_k(\mathbb{S}^3)$ of $\pi_k(\mathbb{S}^3)$ by

$$\mathcal{Q}_k(\mathbb{S}^3) = \{\beta \in \pi_k(\mathbb{S}^3) \mid \langle \iota_3, \beta \rangle = 0\},$$

where $\langle -, - \rangle$ stands for the Samelson product.

The relation (2.2) and the result

$$\langle \iota_3, \beta \circ \delta \rangle = \langle \iota_3, \beta \rangle \circ E^3\delta \ [34, (6.3)] \tag{2.10}$$

lead to:

Lemma 2.6. *The subgroups:*

1. $M_k(\mathbb{S}^{d(n+1)-1}) \subseteq \pi_k(\mathbb{S}^{d(n+1)-1})$;
2. $L'_{k-1}(\mathbb{S}^{d-1}), L''_{k-1}(\mathbb{S}^{d-1}), L_{k-1}(\mathbb{S}^{d-1}) \subseteq \pi_{k-1}(\mathbb{S}^{d-1})$;
3. $\mathcal{Q}_k(\mathbb{S}^3) \subseteq \pi_k(\mathbb{S}^3)$

are admissible.

We set

$$L'^{d-1}_{k-1} = L'_{k-1}(\mathbb{S}^{d-1};2),\ L''^{d-1}_{k-1} = L''_{k-1}(\mathbb{S}^{d-1};2), L^{d-1}_{k-1} = L_{k-1}(\mathbb{S}^{d-1};2),$$
$$P^{d(n+1)-1}_k = P_k(\mathbb{S}^{d(n+1)-1};2), Q^{d-1}_{k-1} = Q_{k-1}(\mathbb{S}^{d-1};2),\ \text{and}$$
$$M^{d(n+1)-1}_k = M_k(\mathbb{S}^{d(n+1)-1};2).$$

The covering $\rho : \mathbb{S}^3 \to SO(3)$ leads to the map $f : \mathbb{S}^3 \times \mathbb{S}^2 \to \mathbb{S}^2$ given by $f(x,y) = \rho(x)y$ for $(x,y) \in \mathbb{S}^3 \times \mathbb{S}^2$. Because $f\ |_{\mathbb{S}^3} = \eta_2$ and $f\ |_{\mathbb{S}^2} = \iota_2$, the results [12], [34, Sect. 9] and [82, p. 115] imply

$$J(\rho) = \nu'^{+} = x\langle\iota_3,\iota_3\rangle \text{ for an integer } 1 \leq x \leq 11 \text{ with } (x,12) = 1. \tag{2.11}$$

Writing $(-,-)$ for the greatest common divisor, we can state:

Lemma 2.7. 1. $L''_{3,n}(\mathbb{S}^3) = \frac{24}{(24,n+1)}\pi_3(\mathbb{S}^3)$, $L''_{4,n}(\mathbb{S}^3) = \pi_4(\mathbb{S}^3)$ *for* $n \geq 1$*;*

2. $$M_{k,\mathbb{F}}(\mathbb{S}^m) = \begin{cases} 0 & \text{for } \mathbb{F} = \mathbb{C},\ k = 7 \text{ and } m = 5,\\ \pi_{12}(\mathbb{S}^9) & \text{for } \mathbb{F} = \mathbb{C}, k = 12 \text{ and } m = 9,\\ \frac{24}{(24,n+1)}\pi_{4n+3}(\mathbb{S}^{4n+3}) & \text{for } \mathbb{F} = \mathbb{H},\ k = 4n+3 \text{ and}\\ & m = 4n+3 \text{ with } n \geq 1,\\ \frac{2}{(2,n+1)}\pi_{4n+6}(\mathbb{S}^{4n+3}) & \text{for } \mathbb{F} = \mathbb{H},\ k = 4n+6 \text{ and}\\ & m = 4n+3 \text{ with } n \geq 1; \end{cases}$$

3. $$Q_k(\mathbb{S}^3) = \begin{cases} 12\pi_3(\mathbb{S}^3) & \text{for } k = 3,\\ 0 & \text{for } k = 4,5,\\ 3\pi_6(\mathbb{S}^3) \cong \mathbb{Z}_4 & \text{for } k = 6. \end{cases}$$

Proof. First, notice that Lemma 2.4(2) and (3) yields:

$$1.\ [\gamma_{2n}, i_{\mathbb{C}}] = \gamma_{2n}\eta_{4n+1} \text{ and } 2.\ [i_{\mathbb{H}}, \gamma_n] = \pm(n+1)\gamma_n\nu^{+}_{4n+3}. \tag{2.12}$$

Then, Lemma 1.2(2), (1.23), and (2.12)(2) lead to $[i_{\mathbb{H}}\eta_4, \gamma_n] = \pm(n+1)\gamma_n\nu_{4n+3}\eta_{4n+6} = 0$, and (1) follows.

By (1.25), (1.18), and (2.12)(1), $[\gamma_2\eta^2_5, i_{\mathbb{C}}] = \gamma_2\eta^3_5 = 4\gamma_2\nu_5 \neq 0$. Hence, $M_{7,\mathbb{C}}(\mathbb{S}^5) = 0$. Because, by (1.23) and (2.12)(1), $[\gamma_4\nu^{+}_9, i_{\mathbb{C}}] = \gamma_4\eta_9\nu^{+}_{10} = 0$, we obtain $M_{12,\mathbb{C}}(\mathbb{S}^9) = \pi_{12}(\mathbb{S}^9)$. Next, Lemma 1.2(2), (1.8), and (2.12)(2) yield $[\gamma_n\nu^{+}_{4n+3}, i_{\mathbb{H}}] = \pm(n+1)\gamma_n\nu^2_{4n+3}$. Consequently, we obtain that $M_{4n+3,\mathbb{C}}(\mathbb{S}^{4n+3}) = \frac{24}{(24,n+1)}\pi_{4n+3}(\mathbb{S}^{4n+3})$, $M_{4n+6,\mathbb{C}}(\mathbb{S}^{4n+3}) = \frac{2}{(2,n+1)}\pi_{4n+6}(\mathbb{S}^{4n+3})$, and (2) is shown.

In view of (2.11), it holds $Q_3(\mathbb{S}^3) = 12\pi_3(\mathbb{S}^3)$. Further, (2.10) and (2.11) lead to $\langle\iota_3,\eta_3\rangle = \langle\iota_3,\iota_3\rangle \circ \eta_6 = \nu'\eta_6 \neq 0$ and $\langle\iota_3,\eta^2_3\rangle = \langle\iota_3,\iota_3\rangle \circ \eta^2_6 = \nu'\eta^2_6 \neq 0$. Thus, we deduce that $Q_4(\mathbb{S}^3) = 0$ and $Q_5(\mathbb{S}^3) = 0$. Finally, (1.25), (2.10), and (2.11) yield $\langle\iota_3,\nu'^{+}\rangle = \langle\iota_3,\iota_3\rangle \circ E^3\nu'^{+} = 2x\nu'^{+} \circ \nu^{+}_6 = \pm\alpha_1(3)\alpha_1(6)$. Then, (1.8) leads to $Q_6(\mathbb{S}^3) = 3\pi_6(\mathbb{S}^3) \cong \mathbb{Z}_4$ and the proof is complete. □

Notice that for $\beta \in \pi_{k-1}(\mathbb{S}^3)$, by Lemma 1.2(2) and (2.12)(2), we get

$$[i_{\mathbb{H}} E\beta, \gamma_n] = \pm(n+1)\gamma_n \nu^+_{4n+3} E^{4n+3}\beta. \tag{2.13}$$

Now, we show:

Lemma 2.8. 1. $L''_{k-1,n}(\mathbb{S}^3) = E^{-4n-3}(\mathrm{Ker}\,(n+1)\nu^+_{4n+3*})$, *where*

$$\nu^+_{4n+3*} : \pi_{k+4n+2}(\mathbb{S}^{4n+6}) \to \pi_{k+4n+2}(\mathbb{S}^{4n+3})$$

is the induced homomorphism. In particular, $L'_{k-1,n}(\mathbb{S}^3) \subseteq L''_{k-1,n}(\mathbb{S}^3)$ *for* n *odd;*

2. $Q_{k-1}(\mathbb{S}^3) = L'_{k-1,n}(\mathbb{S}^3)$ *for* $n \geq 2$*;*
3. *Let* $n \geq 2$*. Then,* $L'_{k-1}(\mathbb{S}^3; 3) \subseteq L''_{k-1}(\mathbb{S}^3; 3)$ *and* $L'_{k-1}(\mathbb{S}^3; p) = L''_{k-1}(\mathbb{S}^3; p) = \pi_{k-1}(\mathbb{S}^3; p)$ *for an odd prime* $p \geq 5$*;*
4. $i_{\mathbb{H}*}P_k(\mathbb{S}^4) \subseteq i_{\mathbb{H}*}EQ_{k-1}(\mathbb{S}^3)$*. Moreover,* $i_{\mathbb{H}*}P_k(\mathbb{S}^4) = i_{\mathbb{H}*}EQ_{k-1}(\mathbb{S}^3)$ *provided that* $2E^4Q_{k-1}(\mathbb{S}^3) = 0$.

Proof. By (2.13), the first half of (1) follows. If $[E\beta, \iota_4] = 0$, then $[E\beta, \nu_4^+] = 0$ (2.2). This leads to the second one of (1) for $n = 1$. The fact that $i_{\mathbb{H}} \circ \nu_4^+ = 0$ for $n \geq 2$ and (1.20) yield to

$$[E\beta, \iota_4] = \pm(2\nu_4^+ - E\nu'^+) \circ E^4\beta, \tag{2.14}$$

the relation $i_{\mathbb{H}}[E\beta, \iota_4] = 0$ implies $i_{\mathbb{H}}E(\nu'^+ E^3\beta) = 0$, and so $\nu'^+ E^3\beta = 0$. Hence,

$$2\nu_5^+ E^5\beta = 0 \tag{2.15}$$

and $(n+1)\nu^+_{4n+3}E^{4n+3}\beta = 0$ for all odd n. This leads to the second one of (1). By means of (1.8) and (2.13), we obtain (2). Given $\beta \in L'_{k-1}(\mathbb{S}^3; 3)$, in view of (2.13) and (2.15), it holds $\beta \in L''_{k-1}(\mathbb{S}^3; 3)$ and the first half of (3) follows. The second one of (3) is a direct consequence of (2.13) and (2.14).

In view of (1.27)(2), we may suppose that $\beta = E\beta_1 + \nu_4^+\beta_2 \in P_k(\mathbb{S}^4)$ for $\beta_1 \in \pi_{k-1}(\mathbb{S}^3)$ and $\beta_2 \in \pi_k(\mathbb{S}^7)$. Then, by the relation $i_{\mathbb{H}}\beta = i_{\mathbb{H}}E\beta_1$, we obtain

$$0 = i_{\mathbb{H}}[\iota_4, \beta] = i_{\mathbb{H}}[\iota_4, E\beta_1] = \pm(i_{\mathbb{H}}E)(\nu'^+ E^3\beta_1).$$

This implies the inclusion $G_k(\mathbb{S}^4) \subseteq EQ_{k-1}(\mathbb{S}^3) \oplus \nu^+_{4\,*}\pi_k(\mathbb{S}^7)$ and leads to the first half of (4). On the other hand, for $\beta \in Q_{k-1}(\mathbb{S}^3)$, we obtain $[\iota_4, E\beta] = 2\nu_4^+ E^4\beta = 0$ by the assumption. This means $EQ_{k-1}(\mathbb{S}^3) \subseteq P_k(\mathbb{S}^4)$ and leads to the second half of (4), and the proof is complete. □

The following has been suggested by K. Morisugi:

Theorem 2.9. $L''_{k-1,n}(\mathbb{S}^3) = \pi_{k-1}(\mathbb{S}^3)$ *for* $n \geq 1$*, if* $5 \leq k \leq 4n+2$.

Proof. Because $J([\iota_3]) = \nu_4^+$, in view of [85, Corollary 11.2], we have

$$\nu_{4n+3}^+ E^{4n+3}\beta = J([\iota_3]_{4n+3}\beta). \tag{2.16}$$

By the fact that $\pi_{k-1}(\mathbb{S}^3)$ is finite for $k \neq 4$ and

$$\pi_{k-1}(SO(4n+3)) \cong \begin{cases} \mathbb{Z} & \text{for } k \equiv 0 \ (\text{mod } 4); \\ \mathbb{Z}_2 & \text{for } k \equiv 1, 2 \ (\text{mod } 8); \\ 0 & \text{for } k \equiv 3, 5, 6, 7 \ (\text{mod } 8) \end{cases}$$

if $k \leq 4n+2$ [14], we have the result for $k \not\equiv 1, 2$ (mod 8).

Assume that $\nu_{4n+3}E^{4n+3}\beta \neq 0$ for $k \equiv 1$ (mod 8) (respectively $k \equiv 2$ (mod 8)). Then, in view of [37, Lemma 2], we have $\nu_{4n+3}E^{4n+3}\beta = J([\iota_3]_{4n+3}\beta) = J(\delta\eta_{k-2})$ (respectively $\nu_{4n+3}E^{4n+3}\beta = J([\iota_3]_{4n+3}\beta) = J(\delta'\eta_{k-3}^2)$), where $\delta \in \pi_{k-2}(SO(4n+3))$ (respectively $\delta' \in \pi_{k-3}(SO(4n+3))$) is a generator. Since $J : \pi_{k-1}(SO(4n+3)) \to \pi_{k+4n+2}(\mathbb{S}^{4n+3})$ is a monomorphism [2, Theorem 1.1] (respectively [2, Theorem 1.3]), we obtain $\delta\eta_{k-2} = [\iota_3]_{4n+3}\beta$ (respectively $\delta'\eta_{k-3}^2 = [\iota_3]_{4n+3}\beta$). But, in view of [16, Theorem 1.1], the element $\delta\eta_{k-2}$ (respectively $\delta'\eta_{k-3}^2$) has the $SO(6)$-of-origin. This contradiction leads to the statement for $k \equiv 1$ (mod 8) (respectively $k \equiv 2$ (mod 8)) and the proof is complete. □

Another Proof of Theorem 2.9. To consider the cases $k \equiv 1, 2$ (mod 8), set $j_{r,n}$ for the generator of $J\pi_r(SO(n))$ with $r \leq n-2$ and write (according to Adams' notations [2]) $j_r = j_{r,\infty}$. First, assume that $\nu_{4n+3}E^{4n+3}\beta \neq 0$ for $k \equiv 1$ (mod 8). Then, in view of [37, Lemma 2], we have $\nu_{4n+3}E^{4n+3}\beta = j_{k-2,4n+3}\eta_{k+4n+1}$ and $j_{k-2,4n+3}\eta_{k+4n+1}^2$ generates $J\pi_k(SO(4n+3))$. On the other hand, by [85, Proposition 3.1],

$$\beta \wedge \eta_2 = E^2(\beta\eta_{k-1}) = \eta_5 E^3\beta \text{ and } \nu_4 \wedge \beta = \nu_7 E^7\beta = (E^4\beta)\nu_{k+3}.$$

Then, (1.23) implies a contradictory relation

$$j_{k-2,4n+3}\eta_{k+4n+1}^2 = \nu_{4n+3}(E^{4n+3}\beta)\eta_{k+4n+2} = \nu_{4n+3}\eta_{4n+6}E^{4n+4}\beta = 0.$$

Now, assume that $\nu_{4n+3}E^{4n+3}\beta \neq 0$ for $5 \leq k \leq 4n+2$ with $k \equiv 2$ (mod 8). Then, $\nu_{4n+3}E^{4n+3}\beta = j_{k-1,4n+3}$ and we get

$$< j_{k-1}, \eta, 2\iota > = < \nu E^\infty\beta, \eta, 2\iota > = < (E^\infty\beta)\nu, \eta, 2\iota >$$

$$\supseteq (E^\infty\beta) < \nu, \eta, 2\iota > \equiv 0 \ (\text{mod } 2\pi_{k+1}^s) \ (1.102)(2)$$

for the stable Toda bracket. On the other hand, by [2, Corollary 11.7] and [63, Theorem B. v)], we deduce that π_{k+1}^s contains the direct summand $\mathbb{Z}_8$ which is generated by $< j_{k-1}, j_3, 24\iota >$ (mod odd components).

Because $j_1 = \eta$, $j_3 = \nu^+$, and $j_{k-1} = j_{k-3} \circ \eta^2$ [37, Lemma 2], we derive

$$< j_{k-3}, j_3, 24\iota > \subseteq < j_{k-3}, 12j_3, 2\iota > = < j_{k-3}, \eta^3, 2\iota > \supseteq < j_{k-1}, \eta, 2\iota > .$$

Thus,

$$< j_{k-3}, j_3, 24\iota > \equiv < j_{k-1}, j_1, 2\iota > \pmod{2\pi^s_{k+1}}.$$

This is a contradiction with the above which leads to the assertion.

We write

$$P'_k(\mathbb{F}P^n) = P_k(\mathbb{F}P^n) \cap \gamma_{n*}\pi_k(\mathbb{S}^{d(n+1)-1})$$

and

$$P''_k(\mathbb{F}P^n) = P_k(\mathbb{F}P^n) \cap i_{\mathbb{F}*}(E\pi_{k-1}(\mathbb{S}^{d-1})).$$

Notice that $P''_k(\mathbb{F}P^n) = 0$ if $d = 1, 2$ and $k \geq d + 1$.

Now, we show:

Proposition 2.10. *$P'_k(\mathbb{F}P^n) = \gamma_{n*}(P_k(\mathbb{S}^{d(n+1)-1}) \cap M_k(\mathbb{S}^{d(n+1)-1}))$ and $P''_k(\mathbb{F}P^n) = i_{\mathbb{F}*}EL_{k-1}(\mathbb{S}^{d-1})$.*

Proof. For $\alpha \in \pi_k(\mathbb{S}^{d(n+1)-1})$, it holds that $[\gamma_n\alpha, \gamma_n] = 0$ if and only if $[\alpha, \iota_{d(n+1)-1}] = 0$. This leads to the first half.

Next, for an element $\beta \in \pi_{k-1}(\mathbb{S}^{d-1})$, it holds that $i_{\mathbb{F}}E\beta \in P''_k(\mathbb{F}P^n)$ if and only if $\beta \in L_{k-1}(\mathbb{S}^{d-1})$. This leads to the second half and completes the proof. □

2.2 Preliminaries II

Since $H((2\iota_6)\bar{\nu}_6)) = 4H(\bar{\nu}_6)$ [85, Proposition 2.2], we derive $(2\iota_6)\bar{\nu}_6 - 4\bar{\nu}_6 = x\varepsilon_6$ for some integer x. Suspending and using (1.67), we get $0 = -2\bar{\nu}_7 = x\varepsilon_7$ and $x \equiv 0 \pmod 2$. Next, by [92, (8.5) Theorem, p. 534], $(2\iota_6)\bar{\nu}_6 = 2\bar{\nu}_6 + [\iota_6, \iota_6]h_0(\bar{\nu}_6)$. Consequently:

$$\begin{aligned} &1.\ (2\iota_6)\bar{\nu}_6 = 4\bar{\nu}_6, \\ &2.\ [\iota_6, \iota_6]h_0(\bar{\nu}_6) = 2\bar{\nu}_6. \end{aligned} \tag{2.17}$$

On the other hand, by [7], it holds $(2\iota_6)\bar{\nu}_6 = 2\bar{\nu}_6 + [\iota_6, \nu_6]$. Hence, (2.17)(1) improves (1.67) as follows:

$$2\bar{\nu}_6 = [\iota_6, \nu_6].$$

In view of [85, Proposition 2.2], it holds $H((2\iota_4)\nu_4^+) = 4H(\nu_4^+)$. Hence, $(2\iota_4)\nu_4^+ - 4\nu_4^+ = xE\nu'^+$ for some integer x. Suspending, we get $-2\nu_5^+ = 2x\nu_5^+$ with $x \equiv -1 \pmod{12}$. Thus, (1.20) yields:

$$\begin{aligned} &1.\ (2\iota_4)\nu_4 = 4\nu_4 - E\nu' = 2\nu_4 \pm [\iota_4, \iota_4], \\ &2.\ (2\iota_4)\nu_4^+ = 4\nu_4^+ - E\nu'^+ = 2\nu_4^+ \pm [\iota_4, \iota_4]. \end{aligned} \tag{2.18}$$

Because $H((2\iota_8)\sigma_8^+) = 4H(\sigma_8^+)$ [85, Proposition 2.2], we get $(2\iota_8)\sigma_8^+ - 4\sigma_8^+ = xE\sigma'^+$ for some integer x. Suspending, we get $-2\sigma_9^+ = 2x\sigma_9^+$ with $x \equiv -1 \pmod{120}$. Further, in view of [92, (8.5) Theorem, p. 534], it holds $(2\iota_8)\sigma_8^+ = 2\sigma_8^+ + [\iota_8, \iota_8]h_0(\sigma_8^+)$. Hence, (1.22) leads to:

$$\begin{aligned} &1.\ (2\iota_8)\sigma_8 = 2\sigma_8 \pm [\iota_8, \iota_8] = 4\sigma_8 - E\sigma', \\ &2.\ (2\iota_8)\sigma_8^+ = 2\sigma_8^+ \pm [\iota_8, \iota_8] = 4\sigma_8^+ - E\sigma'^+, \\ &\quad (3)\ [\iota_8, \iota_8]h_0(\sigma_8^+) = 2\sigma_8^+ - E\sigma'^+. \end{aligned} \tag{2.19}$$

Now, we recall:

$$\nu'\zeta_6 = 0 \text{ [54, Lemma 3.3 (ii)] and [75, Proposition (2.4)(1)]}, \tag{2.20}$$

and

$$\nu'\bar{\varepsilon}_6 \neq 0 \text{ [85, Theorem 12.8]}. \tag{2.21}$$

There is $\bar{\varepsilon}' \in \pi_{20}^3$ such that

$$E\bar{\varepsilon}' = (E\nu')\kappa_7 \text{ [85, Lemma 12.3]}. \tag{2.22}$$

Following [85, p. 103], choose $\rho^{IV} \in \{\sigma''', 2\iota_{12}, 8\sigma_{12}\}_1$, $\rho''' \in \{\sigma'', 4\iota_{13}, 4\sigma_{13}\}_1$, $\rho'' \in \{\sigma', 8\iota_{14}, 2\sigma_{14}\}_1$, and $\rho' \in \{\sigma_9, 16\iota_{16}, \sigma_{16}\}_1$. In the sequel, we write $\rho''^+ = \rho'' + \alpha_4(7) + \alpha_{2,5}(7)$.

There exists $\rho_{13} \in \pi_{28}^{13}$ such that

$$E^4\rho' = 2\rho_{13} \text{ [85, Lemma 10.9]}. \tag{2.23}$$

Denote that $\rho_n = E^{n-13}\rho_{13}$ for $n \geq 13$. Then,

$$\nu_{11}\rho_{14} = 0 \text{ [75, Proposition 2.17(7)] and [85, (12.23)]}. \tag{2.24}$$

By means of [49, Lemma 6.1] and its proof, we have

$$\eta_6\bar{\kappa}_7 \in \{\nu_6^2, 2\iota_{12}, \kappa_{12}\} \ (\text{mod } \nu_6^2 \circ \pi_{27}^{12} + \pi_{13}^6 \circ \kappa_{13} = \nu_6^2 \circ \{E^3\rho', \bar{\varepsilon}_{12}\} + \{\sigma''\} \circ \kappa_{13}).$$

From the fact that $E^\infty(\sigma''\kappa_{13}) = (4\sigma)\kappa = 0$ (1.35)(1), (1.22) and (1.109), $E^\infty(\nu_6^2 E^3\rho') = \nu^2(2\rho) = 0$ (2.23), and $\nu_6^2\bar{\varepsilon}_{12} = \nu_6^2\eta_{12}\kappa_{13} = 0$ (1.108)(1), we have $E^\infty(\pi_{13}^6 \circ \kappa_{13} + \nu_6^2 \circ \pi_{27}^{12}) = 0$. But, in view of [49, Theorem A], it holds $\pi_{27}^6 = \{\eta_6\bar{\kappa}_7\} \cong \mathbb{Z}_2$ and $\pi_{21}^s = \{\eta\bar{\kappa}, \sigma^3\} \cong (\mathbb{Z}_2)^2$. Hence, $E^\infty : \pi_{27}^6 \to \pi_{21}^s$ is a monomorphism and we obtain

$$\eta_6\bar{\kappa}_7 = \{\nu_6^2, 2\iota_{12}, \kappa_{12}\}. \tag{2.25}$$

Next,

$$\eta_7\sigma_8\kappa_{15} = 0 \text{ [49, Lemma 8.1]}, \tag{2.26}$$

$$\varepsilon_5\kappa_{13} = \eta_5^2\bar{\kappa}_7 \text{ [73, I-Proposition 3.1(2)]} \tag{2.27}$$

and $\eta_9\bar{\kappa}_{10} = \bar{\kappa}_9\eta_{29}$ [85, Proposition 3.1] implies

$$\eta_8^2\bar{\kappa}_{10} = (\eta_8\bar{\kappa}_9)\eta_{29}. \tag{2.28}$$

Following [85, p. 153], choose $\nu_{16}^* \in \{\sigma_{16}, 2\sigma_{23}, \nu_{30}\}_1$ and denote $\nu_n^* = E^{n-16}\nu_{16}^*$ for $n \geq 16$. In view of [85, Lemma 12.18], there exists $\lambda \in \pi_{31}^{13}$ such that $H(\lambda) = \nu_{25}^2$ and $E^3\lambda - 2\nu_{16}^* = \pm[\iota_{16}, \nu_{16}]$. Because $2\nu_{16}^*\nu_{34} = 0$ [49, (7.10)], we get $E^3(\lambda\nu_{31}) = [\iota_{16}, \nu_{16}^2]$ and the result [85, Corollary 12.25] implies

$$\nu_{16}\xi_{19} = \nu_{16} \circ ([\iota_{19}, \iota_{19}] - \nu_{19}^*) = [\nu_{16}, \nu_{16}] - \nu_{16}\nu_{19}^* = [\iota_{16}, \nu_{16}^2] - \nu_{16}\nu_{19}^* = E^3(\lambda\nu_{31}) - \nu_{16}\nu_{19}^*.$$

Consequently, (1.50) leads to $\nu_{15}\nu_{18}^* + \nu_{15}\xi_{18} - E^2(\lambda\nu_{31}) \in P\pi_{38}^{31} = P\{\sigma_{31}\} = 0$ and

$$E^2(\lambda\nu_{31}) = \nu_{15}(\nu_{18}^* + \xi_{18}). \tag{2.29}$$

Now, let $\alpha \in \pi_k^5$ and $\beta \in \pi_k^3$ be elements such that $2\iota_5 \circ \alpha = 0$ and $\{\eta_4, 2\iota_5, \alpha\} \ni E\beta$. Then, in view of (1.102)(1), $\nu_6 E^6\beta \in \nu_6 \circ \{\eta_9, 2\iota_{10}, E^5\alpha\} = [\iota_6, \iota_6] \circ E^6\alpha$. Hence,

$$\nu_6 E^6\beta = [\iota_6, \iota_6] \circ E^6\alpha. \tag{2.30}$$

Next, for $\bar{\mu}_3 \in \{\mu_3, 2\iota_{12}, 8\sigma_{12}\}_1$ [85, Chap. XII, (i)], the result [85, Proposition 1.3] and (1.60) imply

$$\bar{\mu}_n \in \{\mu_n, 2\iota_{n+9}, 8\sigma_{n+9}\}_{n-2} \text{ for } n \geq 3. \tag{2.31}$$

Then,

$$\nu_6\bar{\mu}_9 = 16P(\rho_{13}) \text{ [53, (16.6)]}. \tag{2.32}$$

Next, we follow [85, Chap. XII, (iii)] to choose $\xi_{12} \in \{\sigma_{12}, \nu_{19}, \sigma_{22}\}_1$. Then,

$$\sigma_9^3 = \nu_9\xi_{12} \text{ [73, II, Proposition 2.1(2)]}. \tag{2.33}$$

From the fact that $\mu' \in \{\eta_3, 2\iota_4, \mu_4\}_1$, (1.102)(1) and (1.133), we have $\nu_6 E^6 \mu' \in \nu_6 \circ \{\eta_9, 2\iota_{10}, \mu_{10}\} = [\iota_6, \mu_6] = 0$. Hence,

$$\nu_6 E^6 \mu' = 0. \tag{2.34}$$

We recall $\pi_{21}^6 \cong \mathbb{Z}_4 \oplus \mathbb{Z}_2$. Since $\pi_{22}^3 = \{\bar{\mu}', \nu'\mu_6\sigma_{15}\} \cong \mathbb{Z}_4 \oplus \mathbb{Z}_2$, by (2.34) and the definition [85, p. 137] of $\bar{\mu}' \in \{\mu', 4\iota_{14}, 4\sigma_{14}\}_1$, we see that $\nu_6 E^6 \bar{\mu}' \in \circ\nu_6\{E^6\mu, 4\iota_{20}, \sigma_{20}\} \subseteq \{0, 4\iota_{20}, 4\sigma_{20}\} = \pi_{21}^6 \circ 4\sigma_{21} = 0$. This leads to

$$\nu_6 E^6 \bar{\mu}' = 0. \tag{2.35}$$

Next, for an add prime p, consider the exact sequence

$$\pi_i(\mathbb{S}^{2p-1}; p) \xrightarrow{G} \pi_{i+1}(\mathbb{S}^3; p) \xrightarrow{H} \pi_{i+1}(\mathbb{S}^{2p+1}; p) \xrightarrow{\Delta} \pi_{i-1}(\mathbb{S}^{2p-1}; p) \text{ [85, Proposition 13.3]} \tag{2.36}$$

for $i > 2p - 1$. If $p = 3$ and $i = 19$ then all groups in (2.36) are isomorphic to $\mathbb{Z}_3$ and $\pi_{k+18}(\mathbb{S}^{k+5}; 3) = \{\alpha_1(5+k)\beta_1(8+k)\}$ for $k = 0, 2$ [85, Theorem 13.10]. Since $\Delta E^2(\alpha_1(5)\beta_1(8)) = 3(\alpha_1(5)\beta_1(8)) = 0$ (1.9), the map H is an isomorphism and, consequently, G is trivial. Hence,

$$G(\beta) = \alpha_1(3)E\beta = 0 \text{ for } \beta \in \pi_{19}(\mathbb{S}^5; 3). \tag{2.37}$$

Now, we show:

Proposition 2.11. 1. *$\pi_{17}(\mathbb{S}^3; 3) = \{\alpha_1(3)\alpha_3'(6)\} \cong \mathbb{Z}_3$ and $\pi_{17}(\mathbb{S}^3; 5) = \{\alpha_{1,5}(3)\alpha_{1,5}(10)\} \cong \mathbb{Z}_5$;*
2. *$\pi_{19}(\mathbb{S}^3; 3) = \{\alpha_1(3)\alpha_1(6)\beta_1(9)\} \cong \mathbb{Z}_3$;*
3. *$\pi_{21}(\mathbb{S}^3; 3) = \{\alpha_1(3)\alpha_4(6)\} \cong \mathbb{Z}_3$;*
4. *$\pi_{25}(\mathbb{S}^3; 3) = \{\alpha_1(3)\alpha_5(6)\} \cong \mathbb{Z}_3$.*

Proof. We always use the exact sequence in [85, Proposition 13.3]. The first half of (1) follows from (1.10) and [85, Theorem 13.10(ii)], and [85, (13.6)′] leads to the second half of (1). By [85, Theorem 13.10(iv)], we obtain (2). Next, (3) follows from [85, Theorem 13.10(vi)] and [85, Theorem 13.10(vi);(vii)] with the facts $\pi_{24}(\mathbb{S}^3; 3) = 0$, $\pi_{25}(\mathbb{S}^3; 3) \cong \mathbb{Z}_3$ [87, p. 60, Table] yields (4) and the proof is complete. □

Now, we show:

Lemma 2.12. 1. *$M_k(\mathbb{S}^{d(2n+2)-1}) = \pi_k(\mathbb{S}^{d(2n+2)-1})$ for $\mathbb{F} = \mathbb{R}, \mathbb{C}$;*
2. (i) *$E\pi_{k-1}(\mathbb{S}^{2n-1}) \cap \operatorname{Ker} 2\iota_{2n*} \subseteq M_{k,\mathbb{R}}(\mathbb{S}^{2n})$, where $2\iota_{2n*} : \pi_k(\mathbb{S}^{2n}) \to \pi_k(\mathbb{S}^{2n})$ is the induced homomorphism; $M_{k,\mathbb{R}}(\mathbb{S}^{2n}) = \operatorname{Ker} 2\iota_{2n*}$ for $k \leq 4n - 2$;*
   (ii) *$E\pi_{k-1}(\mathbb{S}^{4n}) \cap E^{-1}(\operatorname{Ker} \eta_{4n+1*}) \subseteq M_{k,\mathbb{C}}(\mathbb{S}^{4n+1})$, where $\eta_{4n+1*} : \pi_{k+1}(\mathbb{S}^{4n+2}) \to \pi_{k+1}(\mathbb{S}^{4n+1})$; $M_{k,\mathbb{C}}(\mathbb{S}^{4n+1}) = E^{-1}(\operatorname{Ker} \eta_{4n+1*})$ for $k \leq 8n$;*

(iii) $E\pi_{k-1}(\mathbb{S}^{4n+2}) \cap E^{-3}(\text{Ker } (n+1)\nu^{+}_{4n+3*}) \subseteq M_{k,\mathbb{H}}(\mathbb{S}^{4n+3})$, *where* $\nu^{+}_{4n+3*} : \pi_{k+3}(\mathbb{S}^{4n+6}) \to \pi_{k+3}(\mathbb{S}^{4n+3})$; $M_{k,\mathbb{H}}(\mathbb{S}^{4n+3}) = E^{-3}(\text{Ker } (n+1)\nu^{+}_{4n+3*})$ *for* $k \leq 8n+4$. *In particular,* $M_{k,\mathbb{H}}(\mathbb{S}^{4n+3}) = \pi_k(\mathbb{S}^{4n+3})$ *for* $n+1 \equiv 0 \pmod{24}$;

(iv) $E\pi_{k-1}(\mathbb{S}^{8n+2}) \cap E^{-3}(\text{Ker } (2n+1)\nu^{+}_{8n+3*}) \subseteq M_{k,\mathbb{H}}(\mathbb{S}^{8n+3})$; $M_{k,\mathbb{H}}(\mathbb{S}^{8n+3}) = E^{-3}(\text{Ker } (2n+1)\nu^{+}_{8n+3*})$ *for* $k \leq 16n+4$;

(v) $M_{k,\mathbb{H}}(\mathbb{S}^{8n+7}) = E^{-3}(\text{Ker } 2(n+1)\nu^{+}_{8n+7*})$;

3. $[\iota_{2n}, \iota_{2n}] \in M_{4n-1,\mathbb{R}}(\mathbb{S}^{2n})$;
4. $M_{k,\mathbb{R}}(\mathbb{S}^2) = \pi_k(\mathbb{S}^2)$ *except* $k=2$ *and* $M_{2,\mathbb{R}}(\mathbb{S}^2) = 0$;
5. $G_{k+2n}(\mathbb{S}^{2n}) \subseteq M_{k+2n,\mathbb{R}}(\mathbb{S}^{2n})$ *for* $k \leq 2n-2$;
6. (i) $M_{k,\mathbb{H}}(\mathbb{S}^{8n+3}; p) = G_k(\mathbb{S}^{8n+3}; p) = \pi_k(\mathbb{S}^{8n+3}; p)$ *for an odd prime* $p \geq 5$;
   (ii) $M_{k,\mathbb{H}}(\mathbb{S}^{8n+3}; 3) = \pi_k(\mathbb{S}^{8n+3}; 3)$ *for* $n \equiv 1 \pmod 3$;
7. $$M^{8n+3}_{k+8n+3,\mathbb{H}} = \begin{cases} 8\pi^{8n+3}_{8n+3} & \textit{for } k=0, \\ \pi^{8n+3}_{k+8n+3} & \textit{for } k=1,2,4,5,7,8,9,10, \\ 2\pi^{8n+3}_{k+8n+3} & \textit{for } k=3,6. \end{cases}$$

Proof. (1) is a direct consequence of Lemma 2.4(1);(2).

By Lemma 2.4(1), $E\alpha \in M_{k,\mathbb{R}}(\mathbb{S}^{2n})$ if and only if $2E\alpha = 2\iota_{2n} \circ E\alpha = 0$. This leads to the first half of (2)-(i). The second half is obtained from the Freudenthal suspension theorem. In view of Lemma 2.4(2), $E\alpha \in M_{k,\mathbb{C}}(\mathbb{S}^{4n+1})$ if and only if $\eta_{4n+1}E^2\alpha = 0$ for $\alpha \in \pi_{k-1}(\mathbb{S}^{4n})$. This leads to (2)-(ii).

By means of Lemma 2.4(3), $E\alpha \in M_{k,\mathbb{H}}(\mathbb{S}^{4n+2})$ if and only if $(n+1)\nu^{+}_{4n+3}E^4\alpha = 0$. This leads to (2)-(iii). By the parallel argument, we have (2)-(iv). Since ν_{8n+7} is cyclic (1.31), Lemma 2.4(3) leads to (2)-(v).

(3) is a direct consequence of [24, Proposition 2].

By Lemma 2.4(1), $[\gamma_2, i_{\mathbb{R}}] = -2\gamma_2$ and $[\gamma_2\eta_2, i_{\mathbb{R}}] = 0$. This implies $M_{2,\mathbb{R}}(\mathbb{S}^2) = 0$ and $\eta_2 \in M_{3,\mathbb{R}}(\mathbb{S}^2)$. By Lemma 2.6(1), $\pi_k(\mathbb{S}^2) = \eta_2 \circ \pi_k(\mathbb{S}^3) \subseteq M_{k,\mathbb{R}}(\mathbb{S}^2)$ and this leads to (4).

Suppose that $E\alpha \in P_{2n+k}(\mathbb{S}^{2n})$. Then, $0 = [\iota_{2n}, E\alpha] = [\iota_{2n}, \iota_{2n}] \circ E^{2n}\alpha$. In view of (1.66)(8), we have $0 = H[\iota_{2n}, \iota_{2n}] \circ E^{2n}\alpha = 2E^{2n}\alpha$. Since $E^{2n-1} : \pi_{2n+k}(\mathbb{S}^{2n}) \to \pi_{4n+k-1}(\mathbb{S}^{4n-1})$ is an isomorphism, we obtain $2E\alpha = 0$. Hence, (2)-(i) leads to (5).

Since $s[\iota_{8n+3}, \alpha] = 0$ for $s = 2, p^i$ with some $i \geq 1$ and $t\alpha_1(8n+3)E^3\alpha = 0$ for $t = 3, p^j$ with some $j \geq 1$, we get (6)-(i). The assumptions in (6)-(ii) imply that $2n+1 \equiv 0 \pmod 3$ and $3^i\nu^{+}_{8n+3}E^3\alpha = 3^i[\iota_{8n+3}, \nu_{8n+3}] \circ E^3h_0(\alpha) = 0$ for some $i \geq 1$. Because $24\nu^{+}_{8n+3}E^3\alpha = 2[\iota_{8n+3}, \nu_{8n+3}] \circ E^3h_0(\alpha) = 0$, Lemma 2.4(3) leads to (6)-(ii).

Next, we show (7). By Lemma 2.4(3), we get that $\alpha \in M^{8n+3}_{k+8n+3,\mathbb{H}}$ if and only if $\nu_{8n+3}E^3\alpha = 0$. Hence, $M^{8n+3}_{8n+3,\mathbb{H}} = 8\pi^{8n+3}_{8n+3}$ and in view of the relation (1.23), we get that $M^{8n+3}_{k+8n+3} = \pi^{8n+3}_{8n+k+3}$ for $k = 1, 2$.

Because of (1.19) and (1.34), we derive that $M^{8n+3}_{8n+6,\mathbb{H}} = 2\pi^{8n+3}_{8n+6}$. Evidently, $M^{8n+3}_{8n+7,\mathbb{H}} = M^{8n+3}_{8n+8,\mathbb{H}} = 0$.

The relation $\sharp\nu_{8n+3}^3 = 2$ and (1.34) imply $M_{8n+9,\mathbb{H}}^{8n+3} = 2\pi_{8n+9}^{8n+3} = 0$.

Because $\pi_{8n+10}^{8n+3} = \{\sigma_{8n+3}\} \cong \mathbb{Z}_{16}$, in view of (1.84), we deduce $M_{8n+10,\mathbb{H}}^{8n+3} = \pi_{8n+10}^{8n+3}$.

By means of (1.126), we obtain $\nu_7\varepsilon_{10} = \nu_7\bar{\nu}_{10} = 0$. So, Lemma 2.4(3) leads to $M_{8n+11,\mathbb{H}}^{8n+3} = \pi_{8n+11}^{8n+3}$.

The relations (1.23), (1.39), and (1.72) yield $\nu_{8n+3}\eta_{8n+6}\varepsilon_{8n+7} = 0$, $\nu_{8n+3}^4 = 0$ and $\nu_{8n+3}\mu_{8n+6} = 0$. Hence, again by Lemma 2.4(3), we get $M_{8n+12,\mathbb{H}}^{8n+3} = \pi_{8n+12}^{8n+3}$.

Finally, (1.23) leads to $M_{8n+13,\mathbb{H}}^{8n+3} = \pi_{8n+13}^{8n+3}$ and the proof is completed. □

2.3 Whitehead Center Groups of Projective Spaces

First, notice that by [8, (2.3)] and (2.11), we get $[i_{\mathbb{H}}, i_{\mathbb{H}}] = i_{\mathbb{H}}[\iota_4, \iota_4] = \pm i_{\mathbb{H}}(E\nu'^{+})$ and (2.1) leads to $\sharp[i_{\mathbb{H}}, i_{\mathbb{H}}] = 12$. Writing $[[-,-]]$ for the least common multiple, by Lemma 2.4(3) and the fact that $\sharp[i_{\mathbb{F}}, i_{\mathbb{F}}] = 1$ for $\mathbb{F} = \mathbb{R}, \mathbb{C}$ and $\sharp[i_{\mathbb{H}}, \gamma_n] = \frac{24}{(24,n+1)}$, we have:

Example 2.13. If $n \geq 2$ then:

1. $P_1(\mathbb{R}P^n) = \frac{1+(-1)^{n-1}}{2}\pi_1(\mathbb{R}P^n)$;
2. $P_2(\mathbb{C}P^n) = \frac{3+(-1)^n}{2}\pi_2(\mathbb{C}P^n)$;
3. $P_4(\mathbb{H}P^n) = [[12, \frac{24}{(24,n+1)}]]\pi_4(\mathbb{H}P^n)$.

Next, we show the following result needed in the sequel.

Lemma 2.14. *Let G be an abelian group with $G = G_1 \oplus G_2$ and write $p_k : G \to G_k$ for the projections for $k = 1, 2$. If H is a subgroup of G such that*

$$p_k(H) < H \text{ for } k = 1, 2 \tag{2.38}$$

then $H = G_1 \cap H \oplus G_2 \cap H$.

Proof. Because $p_1(H) < G_1$ and $p_2(H) < G_2$, we have $H = p_1(H) \oplus p_2(H)$. Hence, by the assumption $p_k(H) < H$ for $k = 1, 2$, $p_1(H) < H \cap G_1$ and $p_2(H) < H \cap G_2$. Certainly, $H \cap G_1 < p_1(H)$ and $H \cap G_2 < p_2(H)$. Consequently, $H = H \cap G_1 \oplus H \cap G_2$ and the proof is complete. □

Consider the homomorphisms $[-, i_{\mathbb{F}}] : \pi_k(\mathbb{F}P^n) \to \pi_{k+d-1}(\mathbb{F}P^n)$ and $[-, \gamma_n] : \pi_k(\mathbb{F}P^n) \to \pi_{k+d(n+1)-2}(\mathbb{F}P^n)$. By the bilinearity of the Whitehead products [92, Chap. X, (7.12) Corollary] and (2.2), we obtain

$$P_k(\mathbb{F}P^n) = \operatorname{Ker}\,[-, i_{\mathbb{F}}] \cap \operatorname{Ker}\,[-, \gamma_n].$$

Next, we show:

Lemma 2.15. *Let $n \geq 2$. Then*

1. $\mathrm{Ker}\,[-, i_{\mathbb{F}}] = \gamma_{n*}M_k(\mathbb{S}^{d(n+1)-1}) \oplus i_{\mathbb{F}*}EL'_{k-1}(\mathbb{S}^{d-1})$.
2. *For* $\mathbb{F} = \mathbb{R}, \mathbb{C}$ *and* $k \geq d+1$*:*
   (i) $\mathrm{Ker}\,[-, i_{\mathbb{F}}] = \gamma_{n*}M_k(\mathbb{S}^{d(n+1)-1})$*;*
   (ii) $\mathrm{Ker}\,[-, \gamma_n] = \gamma_{n*}G_k(\mathbb{S}^{d(n+1)-1})$.
3. *For* $\mathbb{F} = \mathbb{H}$*:*
   (i) $(\mathrm{Ker}\,[-, \gamma_n]; p) = \gamma_{n*}\pi_k(\mathbb{S}^{4n+3}; p) \oplus i_{\mathbb{H}*}E\mathrm{L}''_{k-1}(\mathbb{S}^3; p)$ *for an odd prime* p*;*
   (ii) $(\mathrm{Ker}\,[-, \gamma_n]; 2) = \gamma_{n*}G_k(\mathbb{S}^{4n+3}; 2) \oplus i_{\mathbb{H}*}E\mathrm{L}''_{k-1}(\mathbb{S}^3; 2)$ *provided that*

$$[\iota_{4n+3}, \pi_k^{4n+3}] \cap (n+1)\nu_{4n+3}E^{4n+3}\pi_{k-1}^3 = 0. \tag{2.39}$$

Proof. 1. follows from Lemma 2.4 and the relation

$$0 = [\gamma_n\alpha + i_{\mathbb{F}}E\beta, i_{\mathbb{F}}] = [\gamma_n\alpha, i_{\mathbb{F}}] + [i_{\mathbb{F}}E\beta, i_{\mathbb{F}}].$$

In the quaternionic case, this is equivalent to the following two formulas:

$$(n+1)(\nu_{4n+3}^{+}E^3\alpha + [\iota_{4n+3}, \nu_{4n+3}] \circ E^3h_0(\alpha)) = 0 \tag{2.40}$$

and, in view of (1.20),

$$\nu'^{+}E^3\beta = 0. \tag{2.41}$$

By the assumption of (2), $\pi_{k-1}(\mathbb{S}^{d-1}) = 0$. Hence, (2)-(i) follows from (1) and (2)-(ii) is obtained by the fact that $[\alpha, \iota_{d(n+1)-1}] = 0$ if and only if $[\gamma_n\alpha, \gamma_n] = 0$.

For $\gamma_n\alpha + i_{\mathbb{H}}E\beta \in \mathrm{Ker}\,[-, \gamma_n]$, in view of (2.13), we see that

$$\begin{aligned} 0 &= [\gamma_n\alpha + i_{\mathbb{H}}E\beta, \gamma_n] \\ &= \gamma_n[\alpha, \iota_{4n+3}] + [i_{\mathbb{H}}E\beta, \gamma_n] = \gamma_n([\alpha, \iota_{4n+3}] \pm (n+1)\nu_{4n+3}^{+}E^{4n+3}\beta). \end{aligned}$$

That is,

$$[\alpha, \iota_{4n+3}] = \pm(n+1)\nu_{4n+3}^{+}E^{4n+3}\beta. \tag{2.42}$$

Notice that in view of (1.1), Proposition 1.5, and (2.16), under the assumption that the J-homomorphism is a monomorphism, the relations (2.42) are equivalent to

$$\Delta(\alpha) = \pm(n+1)[\iota_3]_{4n+3}\beta.$$

Since $[\iota_{4n+3}, \pi_k(\mathbb{S}^{4n+3}; p)] = 0$ and $P_k(\mathbb{S}^{4n+3}; p) = \pi_k(\mathbb{S}^{4n+3}; p)$ for an odd prime p, (3)-(i) follows. The assumption (2.39) in (3)-(ii) implies that $\alpha \in P_k(\mathbb{S}^{4n+3}; 2)$ and $\beta \in L''_{k-1}(\mathbb{S}^3; 2)$. This leads to (3)-(ii). □

Notice that the condition (2.39) in (3)-(ii) implies the assumption (2.38) of Lemma 2.14.

Now, we obtain:

Proposition 2.16. 1. *Let* $\mathbb{F} = \mathbb{R}, \mathbb{C}$ *and* $k \geq d + 1$. *Then,* $P_k(\mathbb{F}P^n) = \gamma_{n*}(P_k(\mathbb{S}^{d(n+1)-1}) \cap M_k(\mathbb{S}^{d(n+1)-1}))$ *for* $n \geq 2$. *In particular,* $P_k(\mathbb{F}P^{2n+1}) = \gamma_{2n+1*}P_k(\mathbb{S}^{d(2n+2)-1})$ *for* $n \geq 1$;
2. $P_2(\mathbb{R}P^2) = 0$, $P_k(\mathbb{R}P^2) = \pi_k(\mathbb{R}P^2)$ *for* $k \geq 3$ *and* $P_{k+2n}(\mathbb{R}P^{2n}) = \gamma_{2n*}P_{k+2n}(\mathbb{S}^{2n})$ *for* $k \leq 2n - 2$;
3. $P_{k+4n+1}(\mathbb{C}P^{2n}) = \gamma_{2n*}(P_{k+4n+1}(\mathbb{S}^{4n+1}) \cap E^{-1}(\operatorname{Ker} \eta_{4n+1*}))$ *for* $k \leq 4n - 1$.

Proof. By Lemmas 2.12(1) and 2.15(2), we obtain (1).

By (1), Lemma 2.12(1);(4);(5) and the fact that $P_k(\mathbb{S}^2) = \pi_k(\mathbb{S}^2)$ for $k \geq 3$, we obtain (2). By (1) and Lemma 2.12(2)-(ii), we obtain (3). This completes the proof. □

In particular, by means of Lemma 2.12(1) and Example 2.13(2), we derive from Proposition 2.16(1) that $P_k(\mathbb{C}P^3) = \pi_k(\mathbb{C}P^3)$ for all $k \geq 1$ [39].

In the quaternionic case, we obtain:

Proposition 2.17. 1. $P_k(\mathbb{H}P^{2n+1}) = \gamma_{2n+1*}(G_k(\mathbb{S}^{8n+7}) \cap M_k(\mathbb{S}^{8n+7})) \oplus i_{\mathbb{H}*}EQ_{k-1}(\mathbb{S}^3)$. *In particular,* $P_k(\mathbb{H}P^{2n+1}) = \gamma_{2n+1*}G_k(\mathbb{S}^{8n+7}) \oplus i_{\mathbb{H}*}EQ_{k-1}(\mathbb{S}^3)$ *for* $n + 1 \equiv 0 \pmod{12}$;
2. (i) $P_k(\mathbb{H}P^{2n}; 3) = \gamma_{2n*}M_k(\mathbb{S}^{8n+3}; 3) \oplus i_{\mathbb{H}*}EQ_{k-1}(\mathbb{S}^3; 3)$. *In particular,* $P_k(\mathbb{H}P^{2n}; 3) = \gamma_{2n*}\pi_k(\mathbb{S}^{8n+3}; 3) \oplus i_{\mathbb{H}*}EQ_{k-1}(\mathbb{S}^3; 3)$ *for* $n \equiv 1 \pmod 3$;
   (ii) $P_k(\mathbb{H}P^{2n}; p) = \gamma_{2n*}\pi_k(\mathbb{S}^{8n+3}; p) \oplus i_{\mathbb{H}*}E\pi_{k-1}(\mathbb{S}^3; p)$ *for an odd prime* $p \geq 5$;
   (iii) $P_k(\mathbb{H}P^{2n}; 2) = \gamma_{n*}(P_k^{8n+3} \cap M_k^{8n+3}) \oplus i_{\mathbb{H}*}EL^3_{k-1}$ *provided that*

$$[\iota_{8n+3}, M_k^{8n+3}] \cap \nu_{8n+3} \circ E^{8n+3}L'^3_{k-1} = 0; \tag{2.43}$$

3. $P_k(\mathbb{H}P^n) = i_{\mathbb{H}*}EQ_{k-1}(\mathbb{S}^3)$ *for* $5 \leq k \leq 4n + 2$ *and* $n \geq 2$.

Proof. 1. is a direct consequence of Lemmas 2.8(1);(2), 2.12(2)-(iii), and 2.15(1);(3) and the fact that

$$\operatorname{Ker}([-, \gamma_n] \,|_{\operatorname{Ker}[-, i_{\mathbb{H}}]}) = \operatorname{Ker}[-, i_{\mathbb{H}}] \cap \operatorname{Ker}[-, \gamma_n].$$

Let n be even. Then, Lemmas 2.8(3) and 2.15(3)-(i) lead to the first half of (2)-(i). The second half follows from Lemma 2.12(6)-(ii). Lemmas 2.8(3), 2.12(6)-(i), and 2.15(4)-(i) imply (2)-(ii). By Lemma 2.15(1);(3)-(ii), we have (2)-(iii). By Lemmas 2.15(1);(3) and 2.8(2);(6), we obtain (3). This completes the proof. □

Notice that in view of (1.1), Proposition 1.5, and (2.16), under the assumption that the J-homomorphism is a monomorphism, the relation (2.43) is equivalent to

$$\Delta(M_k^{8n+3}) \cap [\iota_3]_{8n+3}L_{k-1}^{\prime 3} = 0. \tag{2.44}$$

Corollary 2.18. *Let* $\alpha \in \pi_k^{4n+3}$ *and* $\beta \in \pi_{k-1}^3$ *satisfy* (2.40) *and* (2.41), *and* $E\alpha' \in M_l^{4n+3}$.

1. *If* $[\alpha, E\alpha'] \neq 0$ *then* $\gamma_n\alpha + i_{\mathbb{H}}E\beta \notin P_k(\mathbb{H}P^n)$;
2. *Let* n *be even. Suppose that* π_k^{4n+3} *is a cyclic group generated by* α *with* $[\alpha, E\alpha'] \neq 0$. *Then, the condition* (2.43) *in Proposition* 2.17(2)-(iii) *holds.*

Proof. First, observe that in view of Lemma 2.4(3), we have $(n+1)\nu_{4n+3}E^4\alpha' = 0$.

1. Suppose that $\gamma_n\alpha + i_{\mathbb{H}}E\beta \in P_k(\mathbb{H}P^n)$. Hence, by (2.42),

$$[\alpha, E\alpha'] = [\alpha, \iota_{4n+3}] \circ E^k\alpha' = \pm(n+1)(E^{4n}\beta) \circ E^{k-4}(\nu_{4n+3}E^4\alpha') = 0.$$

2. Suppose that the condition (2.43) does not hold. Then, $[\alpha, \iota_{4n+3}] = (n+1)\nu_{4n+3}E^{4n+3}\beta$ for some $\beta \in L_{k-1}^{\prime 3}$. Consequently, as in (1),

$$[\alpha, E\alpha'] = [\alpha, \iota_{4n+3}] \circ E^k\alpha' = \pm(n+1)(E^{4n}\beta) \circ E^{k-4}(\nu_{4n+3}E^4\alpha') = 0$$

and this completes the proof. □

2.4 Some Whitehead Center Groups of Real Projective Spaces

In this section, we determine some P-groups of real projective spaces. First, we recall also that

$$\pi_k(\mathbb{S}^3) = \begin{cases} \{\alpha_1(3)\alpha_1(6)\} \cong \mathbb{Z}_3 & \text{for } k = 9, \\ \{\alpha_2(3), \alpha_{1,5}(3)\} \cong \mathbb{Z}_{15} & \text{for } k = 10, \\ \{\varepsilon', \eta_3\mu_4, \alpha_1(3)\alpha_2(6)\} \cong \mathbb{Z}_{12} \oplus \mathbb{Z}_2 & \text{for } k = 13, \\ \{\mu', \varepsilon_3\nu_{11}, \nu'\varepsilon_6, \alpha_3(3), \alpha_{1,7}(3)\} \cong \mathbb{Z}_{84} \oplus (\mathbb{Z}_2)^2 & \text{for } k = 14. \end{cases}$$

Next, we notice that (2.5) yields

$$[\gamma_n\xi \circ E\alpha, i_{\mathbb{R}}] = (-1)^{k+m}[\gamma_n\xi, i_{\mathbb{R}}] \circ E\alpha \tag{2.45}$$

for $\xi \in \pi_{m+1}(\mathbb{S}^n)$, $\alpha \in \pi_k(\mathbb{S}^m)$. In particular, Lemma 2.4(1) leads to

$$[\gamma_{2n}E\alpha, i_{\mathbb{R}}] = (-1)^{k+1}2E\alpha \quad \text{for } \alpha \in \pi_{k-1}(\mathbb{S}^{2n-1}).$$

This implies

$$E\alpha \in M_k(\mathbb{S}^{2n}) \text{ provided that } 2E\alpha = 0 \text{ for } \alpha \in \pi_{k-1}(\mathbb{S}^{2n-1}). \tag{2.46}$$

Now, we are ready to prove:

Theorem 2.19. *Let $n \geq 2$ and $k \leq 13$. Then, the equality $P_{k+n}(\mathbb{R}P^n) = \gamma_{n*}P_{n+k}(\mathbb{S}^n)$ holds except the pairs: $(k,n) = (4,4), (5,4), (6,4), (8,8), (9,8), (10,4), (10,6), (10,8), (10,10), (12,12), (13,12)$, and $(8,13)$.*

Furthermore, $P_8(\mathbb{R}P^4) = \gamma_{4*}\{(E\nu')\eta_7\} \cong \mathbb{Z}_2$, $P_9(\mathbb{R}P^4) = \gamma_{4*}\{(E\nu')\eta_7^2\} \cong \mathbb{Z}_2$ and $P_{10}(\mathbb{R}P^4) = \gamma_{4*}\{\nu_4^2\} \cong \mathbb{Z}_8$, $P_{16}(\mathbb{R}P^8) = \gamma_{8*}\{(E\sigma')\eta_{15}\} \cong \mathbb{Z}_2$, $P_{17}(\mathbb{R}P^8) = \gamma_{8*}\{(E\sigma')\eta_{15}^2\} \cong \mathbb{Z}_2$, $P_{14}(\mathbb{R}P^4) = \gamma_{4*}\{\nu_4^+\sigma'^+ + E\varepsilon', 2E\varepsilon'\} \cong \mathbb{Z}_{120} \oplus \mathbb{Z}_2$, $P_{16}(\mathbb{R}P^6) = \gamma_{6*}\{4\nu_6\sigma_9, \eta_6\mu_7\} \cong (\mathbb{Z}_2)^2$, $P_{18}(\mathbb{R}P^8) = \gamma_{8*}\{x\nu_8^+\sigma_{11}^+ + 2\sigma_8^+\nu_{15}^+\} \cong \mathbb{Z}_{24}$ for x as in (1.82), $P_{20}(\mathbb{R}P^{10}) = \gamma_{10*}\{2\sigma_{10}\nu_{17}, \eta_{10}\mu_{11}\} \cong (\mathbb{Z}_2)^2$, $P_{15}(\mathbb{R}P^4) = \gamma_{4*}\{\nu_4\sigma'\eta_{14}, 2E\mu', \varepsilon_4\nu_{12}, (E\nu')\varepsilon_7\} \cong (\mathbb{Z}_2)^4$, $P_{16}(\mathbb{R}P^4) = \gamma_{4*}\{\nu_4\sigma'\eta_{14}^2, \nu_4^4, (E\nu')\mu_7, (E\nu')\eta_7\varepsilon_8\} \cong (\mathbb{Z}_2)^4$, $P_{24}(\mathbb{R}P^{12}) = \gamma_{12}\{E\theta'\} \cong \mathbb{Z}_2$, $P_{17}(\mathbb{R}P^4) = \gamma_{4*}\{\nu_4^2\sigma_{10}, (E\nu')\eta_7\mu_8\} \cong \mathbb{Z}_8 \oplus \mathbb{Z}_2$, $P_{25}(\mathbb{R}P^{12}) = \gamma_{12*}\{(E\theta')\eta_{24}\} \cong \mathbb{Z}_2$ and $P_{21}(\mathbb{R}P^8) = \gamma_{8*}\{\nu_8\sigma_{11}\nu_{18}\} \cong \mathbb{Z}_2$.

That is:

1. $P_n(\mathbb{R}P^n) = \begin{cases} \pi_n(\mathbb{R}P^n) \cong \mathbb{Z} & \text{for } n = 3, 7, \\ 2\pi_n(\mathbb{R}P^n) \cong 2\mathbb{Z} & \text{for } n \equiv 1 \pmod 2, \text{ unless } n = 3, 7, \\ 0 & \text{for } n \equiv 0 \pmod 2; \end{cases}$

2. $P_{n+1}(\mathbb{R}P^n) = \begin{cases} \pi_3(\mathbb{R}P^2) \cong \mathbb{Z} & \text{for } n = 2, \\ \pi_{n+1}(\mathbb{R}P^n) \cong \mathbb{Z}_2 & \text{for } n = 6 \text{ or } n \equiv 3 \pmod 4, \\ 0 & \text{for otherwise}; \end{cases}$

3. $P_{n+2}(\mathbb{R}P^n) = \begin{cases} \pi_{n+2}(\mathbb{R}P^n) \cong \mathbb{Z}_2 & \text{for } n = 5 \text{ or } n \equiv 2, 3 \pmod 4, \\ 0 & \text{for otherwise}; \end{cases}$

4. $P_{n+3}(\mathbb{R}P^n) = \begin{cases} \pi_5(\mathbb{R}P^2) \cong \mathbb{Z}_2 & \text{for } n = 2, \\ \pi_6(\mathbb{R}P^3) \cong \mathbb{Z}_{12} & \text{for } n = 3, \\ \gamma_{4*}\{3[\iota_4, \iota_4], 2E\nu'\} & \\ \cong 3\mathbb{Z} \oplus \mathbb{Z}_2 & \text{for } n = 4, \\ \pi_{n+3}(\mathbb{R}P^n) \cong \mathbb{Z}_{24} & \text{for } n \equiv 7 \pmod 8 \text{ or} \\ & n = 2^i - 3,\ i \geq 3, \\ 12\pi_{n+3}(\mathbb{R}P^n) \cong \mathbb{Z}_2 & \text{for } n \equiv 1, 3, 5 \pmod 8,\ n \geq 9 \\ & \text{unless } n = 2^i - 3, \\ 2\pi_{n+3}(\mathbb{R}P^n) \cong \mathbb{Z}_{12} & \text{for } n \equiv 2 \pmod 4,\ n \geq 6 \text{ or } n = 12, \\ 0 & \text{for } n \equiv 0 \pmod 4,\ n \geq 8 \text{ unless } n = 12; \end{cases}$

5. $$P_{n+4}(\mathbb{R}P^n) = \begin{cases} \pi_6(\mathbb{R}P^2) \cong \mathbb{Z}_{12} & \textit{for } n = 2, \\ \pi_{n+4}(\mathbb{R}P^n) \cong \mathbb{Z}_2 & \textit{for } n = 3, 5, \\ \gamma_{4*}\{(E\nu')\eta_7\} \cong \mathbb{Z}_2 & \textit{for } n = 4, \\ \pi_{n+4}(\mathbb{R}P^n) = 0 & \textit{for } n \geq 6; \end{cases}$$

6. $$P_{n+5}(\mathbb{R}P^n) = \begin{cases} \pi_{n+5}(\mathbb{R}P^n) \cong \mathbb{Z}_2 & \textit{for } n = 2, 3, 5, \\ \gamma_{4*}\{(E\nu')\eta_7^2\} \cong \mathbb{Z}_2 & \textit{for } n = 4, \\ 3\pi_{11}(\mathbb{R}P^6) \cong 3\mathbb{Z} & \textit{for } n = 6, \\ \pi_{n+5}(\mathbb{R}P^n) = 0 & \textit{for } n \geq 7; \end{cases}$$

7. $$P_{n+6}(\mathbb{R}P^n) = \begin{cases} \pi_{n+6}(\mathbb{R}P^n) \cong \mathbb{Z}_2 & \textit{for } n = 2,\ n \equiv 4, 5, 7 \ (\mathrm{mod}\ 8) \\ & \textit{unless } n = 4 \textit{ or } n = 2^i - 5,\ i \geq 4, \\ \pi_9(\mathbb{R}P^3) \cong \mathbb{Z}_3 & \textit{for } n = 3, \\ \gamma_{4*}\{\nu_4^2\} \cong \mathbb{Z}_8 & \textit{for } n = 4, \\ 0 & \textit{for otherwise}; \end{cases}$$

8. $$P_{n+7}(\mathbb{R}P^n) = \begin{cases} \pi_9(\mathbb{R}P^2) \cong \mathbb{Z}_3 & \textit{for } n = 2, \\ \pi_{10}(\mathbb{R}P^3) \cong \mathbb{Z}_{15} & \textit{for } n = 3, \\ \pi_{12}(\mathbb{R}P^5) \cong \mathbb{Z}_{30} & \textit{for } n = 5, \\ \pi_{14}(\mathbb{R}P^7) \cong \mathbb{Z}_{120} & \textit{for } n = 7, \\ \gamma_{8*}\{3[\iota_8, \iota_8], 4E\sigma'\} & \\ \cong 3\mathbb{Z} \oplus \mathbb{Z}_2 & \textit{for } n = 8, \\ \pi_{n+7}(\mathbb{R}P^n) \cong \mathbb{Z}_{240} & \textit{for } n = 11 \textit{ or } n \equiv 15 \ (\mathrm{mod}\ 16), \\ 120\pi_{n+7}(\mathbb{R}P^n) \cong \mathbb{Z}_2 & \textit{for } n \textit{ odd unless } n = 3, 5, 7, 11 \\ & \textit{and } n \equiv 15 \ (\mathrm{mod}\ 16), \\ 0 & \textit{for } n = 4, 6 \textit{ or } n \textit{ even},\ n \geq 10; \end{cases}$$

9. $$P_{n+8}(\mathbb{R}P^n) = \begin{cases} \pi_{10}(\mathbb{R}P^2) \cong \mathbb{Z}_{15} & \textit{for } n = 2, \\ \pi_{11}(\mathbb{R}P^3) \cong \mathbb{Z}_2 & \textit{for } n = 3, \\ \pi_{14}(\mathbb{R}P^6) \cong \mathbb{Z}_{24} \oplus \mathbb{Z}_2 & \textit{for } n = 6, \\ \pi_{15}(\mathbb{R}P^7) \cong (\mathbb{Z}_2)^3 & \textit{for } n = 7, \\ \gamma_{8*}\{(E\sigma')\eta_{15}\} \cong \mathbb{Z}_2 & \textit{for } n = 8, \\ \gamma_{9*}\{[\iota_9, \iota_9]\} \cong \mathbb{Z}_2 & \textit{for } n = 9, \\ \pi_{n+8}(\mathbb{R}P^n) \cong (\mathbb{Z}_2)^2 & \textit{for } n{=}10 \textit{ or } n \equiv 3 \ (\mathrm{mod}\ 4),\ n \geq 11, \\ 0 & \textit{for } n \equiv 0, 1 \ (\mathrm{mod}\ 4),\ n \geq 4 \\ & \textit{unless } n = 8, 9 \textit{ or} \\ & n \equiv 22 \ (\mathrm{mod}\ 32),\ n \geq 54, \\ \gamma_{n*}\{\varepsilon_n\} \cong \mathbb{Z}_2 & \textit{for } n \equiv 2 \ (\mathrm{mod}\ 8),\ n \geq 18, \\ \gamma_{n*}\{\eta_n\sigma_{n+1}\} \cong \mathbb{Z}_2 & \textit{for } n = 22,\ n \equiv 14 \ (\mathrm{mod}\ 16) \\ & \textit{or } n \equiv 6 \ (\mathrm{mod}\ 32),\ n \geq 14; \end{cases}$$

10. $P_{n+9}(\mathbb{R}P^n) = \begin{cases} \pi_{11}(\mathbb{R}P^2) \cong \mathbb{Z}_2 & \textit{for } n = 2, \\ \pi_{n+9}(\mathbb{R}P^n) \cong (\mathbb{Z}_2)^3 & \textit{for } n = 3, 6, \\ \pi_{16}(\mathbb{R}P^7) \cong (\mathbb{Z}_2)^4 & \textit{for } n = 7, \\ \pi_{n+9}(\mathbb{R}P^n) \cong (\mathbb{Z}_2)^3 & \textit{for } n \equiv 3 \pmod 4,\ n \geq 11, \\ \gamma_{8*}\{(E\sigma')\eta_{15}^2\} \cong \mathbb{Z}_2 & \textit{for } n = 8, \\ \gamma_{9*}\{\sigma_9\eta_{16}^2, \nu_9^3, \eta_9\varepsilon_{10}\} \cong (\mathbb{Z}_2)^3 & \textit{for } n = 9, \\ \gamma_{10*}\{3[\iota_{10}, \iota_{10}], \nu_{10}^3, \eta_{10}\varepsilon_{11}\} & \\ \cong 3\mathbb{Z} \oplus (\mathbb{Z}_2)^2 & \textit{for } n = 10, \\ \gamma_{n*}\{\eta_n\varepsilon_{n+1}\} \cong \mathbb{Z}_2 & \textit{for } n \equiv 1 \pmod 8,\ n \geq 17 \\ & \textit{unless } n = 2^i - 7, \\ 0 & \textit{for } n \equiv 0 \pmod 8,\ n \geq 16, \\ \gamma_{n*}\{\nu_n^3, \eta_n\varepsilon_{n+1}\} \cong (\mathbb{Z}_2)^2 & \textit{for } n \equiv 2 \pmod 4,\ n \geq 14, \\ & n = 2^i - 7,\ i \geq 5 \\ & \textit{or } n \equiv 5 \pmod 8 \\ & \textit{unless } n \equiv 53 \pmod{64}, \\ \gamma_{n*}\{\nu_n^3\} \cong \mathbb{Z}_2 & \textit{for } n \equiv 4 \pmod 8 \textit{ or} \\ & n \equiv 53 \pmod{64},\ n \geq 117; \end{cases}$

11. $P_{n+10}(\mathbb{R}P^n) = \begin{cases} \pi_{12}(\mathbb{R}P^2) \cong (\mathbb{Z}_2)^2 & \textit{for } n = 2, \\ \pi_{13}(\mathbb{R}P^3) \cong \mathbb{Z}_{12} \oplus \mathbb{Z}_2 & \textit{for } n = 3, \\ \gamma_{4*}\{\nu_4^+\sigma'^+ + E\varepsilon', 2E\varepsilon'\} \cong \mathbb{Z}_{120} \oplus \mathbb{Z}_2 & \textit{for } n = 4, \\ \pi_{15}(\mathbb{R}P^5) \cong \mathbb{Z}_{72} \oplus \mathbb{Z}_2 & \textit{for } n = 5, \\ \gamma_{6*}\{4\nu_6\sigma_9, \eta_6\mu_7\} \cong (\mathbb{Z}_2)^2 & \textit{for } n = 6, \\ \pi_{17}(\mathbb{R}P^7) \cong \mathbb{Z}_{24} \oplus \mathbb{Z}_2 & \textit{for } n = 7, \\ \gamma_{8*}\{x\nu_8^+\sigma_{11}^+ + 2\sigma_8^+\nu_{15}^+\} \cong \mathbb{Z}_{24} & \textit{for } n = 8 \textit{ and } x \textit{ as in } (1.82), \\ \gamma_{9*}\{\sigma_9\nu_{16}, \beta_1(9)\} \cong \mathbb{Z}_{24} & \textit{for } n = 9, \\ \gamma_{10*}\{2\sigma_{10}\nu_{17}, \eta_{10}\mu_{11}\} \cong (\mathbb{Z}_2)^2 & \textit{for } n = 10, \\ \pi_{21}(\mathbb{R}P^{11}) \cong \mathbb{Z}_6 \oplus \mathbb{Z}_2 & \textit{for } n = 11, \\ \pi_{n+10}(\mathbb{R}P^n) \cong \mathbb{Z}_6 & \textit{for } n \equiv 3 \pmod 4,\ n \geq 15, \\ 0 & \textit{for } n \equiv 0 \pmod 4,\ n \geq 12, \\ \pi_{n+10}(\mathbb{R}P^n; 2) \cong \mathbb{Z}_2 & \textit{for } n \equiv 2 \pmod 4,\ n \geq 14, \\ \gamma_{n*}\{\beta_1(n)\} \cong \mathbb{Z}_3 & \textit{for } n \equiv 1 \pmod 4,\ n \geq 13; \end{cases}$

12. $P_{n+11}(\mathbb{R}P^n) = \begin{cases} \pi_{13}(\mathbb{R}P^2) \cong \mathbb{Z}_{12} \oplus \mathbb{Z}_2 & \textit{for } n = 2, \\ \pi_{14}(\mathbb{R}P^3) \cong \mathbb{Z}_{84} \oplus (\mathbb{Z}_2)^2 & \textit{for } n = 3, \\ \gamma_{4*}\{\nu_4\sigma'\eta_{14}, 2E\mu', \varepsilon_4\nu_{12}, (E\nu')\varepsilon_7\} & \\ \cong (\mathbb{Z}_2)^4 & \textit{for } n = 4, \\ \pi_{16}(\mathbb{R}P^5) \cong \mathbb{Z}_{504} \oplus (\mathbb{Z}_2)^2 & \textit{for } n = 5, \end{cases}$

$$13.\ P_{n+11}(\mathbb{R}P^n) = \begin{cases} \gamma_{6*}\{\bar{\nu}_6\nu_{14}, 4\zeta_6\} \cong \mathbb{Z}_4 \oplus \mathbb{Z}_2 & \textit{for } n = 6,\\ \pi_{n+11}(\mathbb{R}P^n) \cong \mathbb{Z}_{504} \oplus \mathbb{Z}_2 & \textit{for } n = 7, 9,\\ \gamma_{8*}\{\bar{\nu}_8\nu_{16}\} \cong \mathbb{Z}_2 & \textit{for } n = 8,\\ \gamma_{12*}\{3[\iota_{12}, \iota_{12}]\} \cong 3\mathbb{Z} & \textit{for } n = 12,\\ \pi_{n+11}(\mathbb{R}P^n) \cong \mathbb{Z}_{504} & \textit{for } n \equiv 1 \pmod 2,\ n \geq 11\\ & \textit{unless } n \equiv 115 \pmod{128},\\ 0 & \textit{for } n \equiv 0 \pmod 4,\ n \geq 16,\\ 2\pi_{n+11}(\mathbb{R}P^n) \cong \mathbb{Z}_{252} & \textit{for } n \equiv 115 \pmod{128},\ n \geq 243,\\ 4\gamma_{n*}\pi^n_{n+11} \cong \mathbb{Z}_2 & \textit{for } n \equiv 2 \pmod 4,\ n \geq 10; \end{cases}$$

$$14.\ P_{n+12}(\mathbb{R}P^n) = \begin{cases} \pi_{14}(\mathbb{R}P^2) \cong \mathbb{Z}_{84} \oplus (\mathbb{Z}_2)^2 & \textit{for } n = 2,\\ \pi_{15}(\mathbb{R}P^3) \cong (\mathbb{Z}_2)^2 & \textit{for } n = 3,\\ \gamma_{4*}\{\nu_4\sigma'\eta^2_{14}, \nu^4_4, (E\nu')\mu_7, & \\ (E\nu')\eta_7\varepsilon_8\} \cong (\mathbb{Z}_2)^4 & \textit{for } n = 4,\\ \pi_{17}(\mathbb{R}P^5; 2) \cong (\mathbb{Z}_2)^3 & \textit{for } n = 5,\\ \pi_{18}(\mathbb{R}P^6) \cong \mathbb{Z}_{240} & \textit{for } n = 6,\\ \pi_{n+12}(\mathbb{R}P^n) = 0 & \textit{for } n = 7, 8, 9,\\ \gamma_{10*}\pi^{10}_{22} \cong \mathbb{Z}_4 & \textit{for } n = 10,\\ \gamma_{12*}\{E\theta'\} \cong \mathbb{Z}_2 & \textit{for } n = 12,\\ \pi_{n+12}(\mathbb{R}P^n) \cong \mathbb{Z}_2 & \textit{for } n = 11, 13,\\ \pi_{n+12}(\mathbb{R}P^n) = 0 & \textit{for } n \geq 14; \end{cases}$$

$$15.\ P_{n+13}(\mathbb{R}P^n) = \begin{cases} \pi_{15}(\mathbb{R}P^2) \cong (\mathbb{Z}_2)^2 & \textit{for } n = 2,\\ \pi_{n+13}(\mathbb{R}P^n) \cong \mathbb{Z}_6 & \textit{for } n = 3, 7, 9,\\ P_{17}(\mathbb{R}P^4) = \gamma_{4*}\{\nu^2_4\sigma_{10}, & \\ (E\nu')\eta_7\mu_8\} \cong \mathbb{Z}_8 \oplus \mathbb{Z}_2 & \textit{for } n = 4,\\ \pi_{n+13}(\mathbb{R}P^n) \cong \mathbb{Z}_6 \oplus \mathbb{Z}_2 & \textit{for } n = 5, 11,\\ \pi_{n+13}(\mathbb{R}P^n) \cong \mathbb{Z}_2 & \textit{for } n = 6, 10,\\ \gamma_{8*}\{\nu_8\sigma_{11}\nu_{18}\} \cong \mathbb{Z}_2 & \textit{for } n = 8,\\ \gamma_{12*}\{(E\theta')\eta_{24}\} \cong \mathbb{Z}_2 & \textit{for } n = 12,\\ \pi_{26}(\mathbb{R}P^{13}) \cong \mathbb{Z}_6 & \textit{for } n = 13,\\ \gamma_{14*}\{3[\iota_{14}, \iota_{14}]\} \cong 3\mathbb{Z} & \textit{for } n = 14,\\ \pi_{n+13}(\mathbb{R}P^n) \cong \mathbb{Z}_3 & \textit{for } n \textit{ odd},\ n \geq 15,\\ 0 & \textit{for } n \textit{ even},\ n \geq 16. \end{cases}$$

Proof. For n odd, the required equality holds by Proposition 2.16(1). For $n = 2$ or the case $k \leq n - 2$ with n even, the equality holds by Proposition 2.16(2).

Let $k \leq 7$. Then, it suffices to prove for the following pairs (k, n):

(i) $(3, 4)$. In virtue of Lemma 2.12(3), $3[\iota_4, \iota_4] \in M_7(\mathbb{S}^4)$ and by (2.46), $[\gamma_4 E\nu', i_{\mathbb{R}}] = -2\gamma_4 E\nu'$. So, $P_7(\mathbb{S}^4) = \{3[\iota_4, \iota_4], 2E\nu'\} \subseteq M_7(\mathbb{S}^4)$. Then, Proposition 2.16(1) leads to the equality $P_7(\mathbb{R}P^4) = \gamma_{4*}P_7(\mathbb{S}^4)$.

(ii) $(4, 4)$. Recall that $P_8(\mathbb{S}^4) = \pi_8(\mathbb{S}^4) = \{\nu_4\eta_7, (E\nu')\eta_7\}$ Sect. 1.2.

The result [92, (8.5) Theorem, p. 534] yields $(2\iota_4)\nu^+_4 = 2\nu^+_4 + [\iota_4, \iota_4]h_0(\nu^+_4)$. Then, (2.18)(2) leads to

$$[\iota_4, \iota_4]h_0(\nu_4^+) = 2\nu_4^+ - E\nu'^+ \tag{2.47}$$

and, by Lemma 2.4(1), we obtain

$$[\gamma_4\nu_4^+, i_{\mathbb{R}}] = -\gamma_4 E\nu'^+. \tag{2.48}$$

Observe that this yields

$$[\gamma_4\nu_4, i_{\mathbb{R}}] = -\gamma_4 E\nu'. \tag{2.49}$$

Hence, in view of (2.45), we get $[\gamma_4\nu_4\eta_7, i_{\mathbb{R}}] = \gamma_4(E\nu')\eta_7 \neq 0$. Further, (2.46) yields $[\gamma_4(E\nu')\eta_7, i_{\mathbb{R}}] = 0$. Thus, the above implies $M_8(\mathbb{S}^4) = \{(E\nu')\eta_7\}$ and, by Proposition 2.16(1), we get $P_8(\mathbb{R}P^4) = \gamma_{4*}\{(E\nu')\eta_7\}$.

(iii) $(5,4), (5,6)$. Recall from Sect. 1.2 that

$$G_{n+5}(\mathbb{S}^n) = \begin{cases} \pi_9(\mathbb{S}^4) = \{\nu_4\eta_7^2, (E\nu')\eta_7^2\} & \text{for } n = 4, \\ 3\pi_{11}(\mathbb{S}^6) = \{3[\iota_6, \iota_6]\} & \text{for } n = 6. \end{cases}$$

The argument parallel to the above leads to $P_9(\mathbb{R}P^4) = \gamma_{4*}\{(E\nu')\eta_7^2\}$.

By Lemma 2.12(3), $M_{11}(\mathbb{S}^6) = \pi_{11}(\mathbb{S}^6)$. So, in view of Proposition 2.16(1), we obtain $P_{11}(\mathbb{R}P^6) = \gamma_{6*}P_{11}(\mathbb{S}^6) = 3\pi_{11}(\mathbb{R}P^6)$.

(iv) $(6,4), (6,6)$. Recall from Sect. 1.2 that

$$G_{n+6}(\mathbb{S}^n) = \begin{cases} \pi_{10}(\mathbb{S}^4) = \{(\nu_4^+)^2, \alpha_1(4)\alpha_1(7)\} & \text{for } n = 4, \\ 0 & \text{for } n = 6. \end{cases}$$

The relations (1.28), (2.45) and (2.48) lead to

$$[\gamma_4(\nu_4^+)^2, i_{\mathbb{R}}] = -2\gamma_4\alpha_1(4)\alpha_1(7). \tag{2.50}$$

Next, (2.46) leads to

$$[\gamma_4\alpha_1(4)\alpha_1(7), i_{\mathbb{R}}] = -2\gamma_4\alpha_1(4)\alpha_1(7).$$

This implies $[\gamma_4((\nu_4^+)^2 - \alpha_1(4)\alpha_1(7)), i_{\mathbb{R}}] = 0$ and Proposition 2.16(1) yields $P_{10}(\mathbb{R}P^4) = \gamma_{4*}\{\nu_4^2\}$.

Because $G_{12}(\mathbb{S}^6) = 0$, Proposition 2.16(1) leads to $P_{12}(\mathbb{R}P^6) = 0$.

(v) $(7,6), (7,8)$. Recall from Sect. 1.2 that

$$G_{n+7}(\mathbb{S}^n) = \begin{cases} 0 & \text{for } n = 6, \\ \{3[\iota_8, \iota_8], 4E\sigma'\} & \text{for } n = 8. \end{cases}$$

Because $G_{13}(\mathbb{S}^6) = 0$, Proposition 2.16(1) leads to $P_{13}(\mathbb{R}P^6) = 0$.

By Lemma 2.12(3) and (2.46) we get $P_{15}(\mathbb{S}^8) \subseteq M_{15}(\mathbb{S}^8)$. Then, Proposition 2.16(1) leads to the equality $P_{15}(\mathbb{R}P^8) = \gamma_{8*}P_{15}(\mathbb{S}^8)$.

Let $8 \leq k \leq 13$. Then, it is sufficient to check for the pairs (k, n) with n even:

(i) $(8,4), (8,6), (8,8)$. Recall from Sect. 1.6 that:

$$G_{n+8}(\mathbb{S}^n) = \begin{cases} 0 & \text{for } n = 4, \\ \pi_{14}(\mathbb{S}^6) = \{\bar{\nu}_6, \varepsilon_6, [\iota_6, \iota_6]\alpha_1(11)\} & \text{for } n = 6, \\ \{(E\sigma')\eta_{15}, \sigma_8\eta_{15} + \bar{\nu}_8 + \varepsilon_8\} & \text{for } n = 8. \end{cases}$$

Certainly, $P_{12}(\mathbb{R}P^4) = \gamma_{4*}G_{12}(\mathbb{S}^4)$. Notice that (2.46) leads to $\varepsilon_6 \in M_{14}(\mathbb{S}^6)$ and, in view of Lemma 2.12(3) and (2.45), we get $[\iota_6, \iota_6]\alpha_1(11) \in M_{14}(\mathbb{S}^6)$.

In view of Lemma 2.4(1) and (2.17)(2), we get

$$[\gamma_6\bar{\nu}_6, i_{\mathbb{R}}] = 0. \tag{2.51}$$

Consequently, $P_{14}(\mathbb{R}P^6) = \gamma_{6*}G_{14}(\mathbb{S}^6)$.

Next, Lemma 2.4(1) and (2.19)(3) lead to

$$[\gamma_8\sigma_8^+, i_{\mathbb{R}}] = \gamma_8 E\sigma'^+ \tag{2.52}$$

which implies

$$[\gamma_8\sigma_8, i_{\mathbb{R}}] = \gamma_8 E\sigma'. \tag{2.53}$$

Then, in view of (2.45), $[\gamma_8\sigma_8\eta_{15}, i_{\mathbb{R}}] = \gamma_8(E\sigma')\eta_{15} \neq 0$, and, by (2.46), it holds $[\gamma_8(E\sigma')\eta_{15}, i_{\mathbb{R}}] = 0$. Thus, we get $P_{16}(\mathbb{R}P^8) = \gamma_{8*}\{(E\sigma')\eta_{15}\}$.

(ii) $(9,4), (9,6), (9,8), (9,10)$. Recall from Sect. 1.6 that:

$$G_{n+9}(\mathbb{S}^n) = \begin{cases} \{\nu_4^3\} & \text{for } n = 4, \\ \pi_{15}(\mathbb{S}^6) & \text{for } n = 6, \\ \{(E\sigma')\eta_{15}^2, \sigma_8\eta_{15}^2 + \nu_8^3 + \eta_8\varepsilon_9\} & \text{for } n = 8, \\ \{3[\iota_{10}, \iota_{10}], \nu_{10}^3, \eta_{10}\varepsilon_{11}\} & \text{for } n = 10. \end{cases}$$

Observe that $P_{n+9}(\mathbb{R}P^n) = \gamma_{n*}G_{n+9}(\mathbb{S}^n)$ for $n = 4, 6$. Further, (2.46) yields $(E\sigma')\eta_{15}^2 \in M_{17}(\mathbb{S}^8)$ and, in view of (2.45), (2.46), and (2.52), we get $[\gamma_8(\sigma_8\eta_{15}^2 + \nu_8^3 + \eta_8\varepsilon_9), i_{\mathbb{R}}] = \gamma_8(E\sigma')\eta_{15}^2 \neq 0$. Hence, $P_{17}(\mathbb{R}P^8) = \gamma_{8*}\{(E\sigma')\eta_{15}^2\}$. Next, (2.46) leads to $\nu_{10}^3, \eta_{10}\varepsilon_{11} \in M_{19}(\mathbb{S}^{10})$ and, by means of Lemma 2.12(3), we get $[\iota_{10}, \iota_{10}] \in M_{19}(\mathbb{S}^{10})$. Thus, $P_{19}(\mathbb{R}P^{10}) = \gamma_{10*}G_{19}(\mathbb{S}^{10})$ as well.

(iii) $(10,4), (10,6), (10,8), (10,10)$. Recall from Sect. 1.6 that:

$$G_{n+10}(\mathbb{S}^n) = \begin{cases} \{\nu_4^+\sigma'^+ \pm E\varepsilon', 2E\varepsilon', \alpha_1(4)\alpha_2(7), \\ \nu_4\alpha_2(7), \nu_4\alpha_1'(7)\}, & \text{if } n = 4; \\ \{\nu_6\sigma_9, \eta_6\mu_7, 3\beta_1(6)\} & \text{for } n = 6; \\ \{\sigma_8\nu_{15}, \nu_8\sigma_{11}, \sigma_8^+\alpha_1(15)\} & \text{for } n = 8; \\ \pi_{20}^{10} = \{\sigma_{10}\nu_{17}, \eta_{10}\mu_{11}\} & \text{for } n = 10. \end{cases}$$

Certainly, $[2\gamma_4 E\varepsilon', i_{\mathbb{R}}] = 0$ and $[\gamma_4\nu_4^+\alpha_{1,5}(7), i_{\mathbb{R}}] = -\gamma_4(E\nu'^+)\alpha_{1,5}(7) = 0$. Further, (2.46) yields $[\gamma_4\alpha_1(4)\alpha_2(7), i_{\mathbb{R}}] = -2\gamma_4\alpha_1(4)\alpha_2(7)$ and, in view of (2.45) and (2.48), it holds $[\gamma_4\nu_4^+\alpha_2(7), i_{\mathbb{R}}] = -\gamma_4(E\nu'^+)\alpha_2(7) = -2\gamma_4\alpha_1(4)\alpha_2(7)$. Hence, $[\gamma_4(\alpha_1(4)\alpha_2(7) - \nu_4^+\alpha_2(7)), i_{\mathbb{R}}] = 0$.

Next, in view of (1.66)(1),(2) and (2.7), it holds $H(\nu_4) = h_0(\nu_4) = \iota_7$ and $H(\sigma') = h_0(\sigma') = \eta_{13}$. Then, the monomorphism $E : \pi_{14}(\mathbb{S}^7) \to \pi_{15}(\mathbb{S}^8)$, (1.23) and Lemma 2.5 lead to $h_0(\nu_4\sigma') = \sigma'$. Consequently, (1.20) yields

$$[\iota_4, \iota_4]h_0(\nu_4\sigma') = (2\nu_4 - E\nu')\sigma'. \tag{2.54}$$

The proof of [92, Theorem (8.6), p. 534] yields that the Hopf–Hilton invariants $h_j(\sigma') \in \pi_{14}(\mathbb{S}^{6k+1}) = 0$ with $k \geq 3$ for $j \geq 1$. Thus, $2h_0(\sigma') = 0$ and [92, (8.6) Theorem, p. 534] imply

$$(2\nu_4 - E\nu')\sigma' = 2\nu_4\sigma' - (E\nu')\sigma' - (2[\nu_4, E\nu'])h_0(\sigma') = 2\nu_4\sigma' - (E\nu')\sigma'.$$

Hence, the relation $(E\nu')\sigma' = 2E\varepsilon' = \eta_4^2\varepsilon_6$ [54, (3.2)] and (2.54) yield

$$[\iota_4, \iota_4]h_0(\nu_4\sigma') = 2\nu_4\sigma' - 2E\varepsilon'.$$

Then, Lemma 2.4(1) leads to

$$[\gamma_4\nu_4\sigma', i_{\mathbb{R}}] = -2\gamma_4 E\varepsilon' \tag{2.55}$$

and

$$[\gamma_4(\nu_4\sigma' + E\varepsilon'), i_{\mathbb{R}}] = \gamma_4(-2\nu_4\sigma' - 2E\varepsilon' + 2\nu_4\sigma' - 2E\varepsilon') = 0.$$

Consequently, in view of the presentations of σ'^+ and ν_4^+, we get $P_{14}(\mathbb{R}P^4) = \gamma_{4*}\{\nu_4\sigma' + E\varepsilon', 2E\varepsilon', \alpha_1(4)\alpha_2(7) - \nu_4^+\alpha_2(7), \nu_4^+\alpha_{1,5}(7)\} = \gamma_{4*}\{\nu_4^+\sigma'^+ + E\varepsilon', 2E\varepsilon'\}$.

Notice that $[\gamma_6\nu_6\sigma_9, i_{\mathbb{R}}] = -2\gamma_6\nu_6\sigma_9$ and $[3\gamma_6\beta_1(6), i_{\mathbb{R}}] = -6\gamma_6\beta_1(6) \neq 0$. Hence, (2.46) yields $P_{16}(\mathbb{R}P^6) = \gamma_{6*}\{4\nu_6\sigma_9, \eta_6\mu_7\}$.

By (1.9), (2.45), (2.52), and [85, Proposition 3.1], we get that $[\gamma_8\sigma_8^+\alpha_1(15), i_{\mathbb{R}}] = \gamma_8(E\sigma'^+)\alpha_1(15) = \gamma_8\alpha_2(8)\alpha_1(15) = \gamma_8\alpha_1(8)\alpha_2(11) = 0$. Next, in view of Lemma 2.12(3), $[\gamma_8(\nu_8\sigma_{11} - 2x\sigma_8\nu_{15}), i_{\mathbb{R}}] = 0$ for x as in (1.82). Thus, we have $P_{18}(\mathbb{R}P^8) = \gamma_{8*}\{\nu_8\sigma_{11} - 2x\sigma_8\nu_{15} + \sigma_8^+\alpha_1(15)\}$. One can also easily observe that $P_{18}(\mathbb{R}P^8) = \gamma_{8*}\{x\nu_8^+\sigma_{11}^+ + 2\sigma_8^+\nu_{15}^+\}$ for x as in (1.82).

Further, the relation $[\gamma_{10}\sigma_{10}\nu_{17}, i_{\mathbb{R}}] = -2\gamma_{10}\sigma_{10}\nu_{17}$ and (2.46) lead to $P_{20}(\mathbb{R}P^{10}) = \gamma_{10*}\{2\sigma_{10}\nu_{17}, \eta_{10}\mu_{11}\}$.

(iv) $(11,4), (11,6), (11,8), (11,10), (11,12)$. Recall from Sect. 1.6 that:

$$G_{n+11}(\mathbb{S}^n) = \begin{cases} \{\nu_4\sigma'\eta_{14}, \nu_4\bar{\nu}_7, \nu_4\varepsilon_7, \\ 2E\mu', \varepsilon_4\nu_{12}, (E\nu')\varepsilon_7\} & \text{for } n = 4; \\ \{\bar{\nu}_6\nu_{14}, 4\zeta_6\} & \text{for } n = 6; \\ \{\bar{\nu}_8\nu_{16}\} & \text{for } n = 8; \\ 4\pi_{21}^{10} = \{4\zeta_{10}\} & \text{for } n = 10; \\ \{3[\iota_{12}, \iota_{12}]\} & \text{for } n = 12. \end{cases}$$

Certainly, by (2.46), it holds $P_{n+11}(\mathbb{R}P^n) = \gamma_{n*}G_{n+11}(\mathbb{S}^n)$ for $n = 8, 10$.

In view of (2.45) and (2.55), we get $[\gamma_4\nu_4\sigma'\eta_{14}, i_{\mathbb{R}}] = -2\gamma_4(E\varepsilon')\eta_{14} = 0$. The relations (1.122), (2.45), (2.49), and [85, Theorem 7.4] lead to $[\gamma_4\nu_4\bar{\nu}_7, i_{\mathbb{R}}] = -\gamma_4(E\nu')\bar{\nu}_7 = \gamma_4\varepsilon_4\nu_{12} \neq 0$ and $[\gamma_4\nu_4\varepsilon_7, i_{\mathbb{R}}] = -\gamma_4(E\nu')\varepsilon_7 \neq 0$. Because (2.46) leads to $[2\gamma_4 E\mu', i_{\mathbb{R}}] = [\gamma_4\varepsilon_4\nu_{12}, i_{\mathbb{R}}] = [\gamma_4(E\nu')\varepsilon_7, i_{\mathbb{R}}] = 0$, we deduce that

$$P_{15}(\mathbb{R}P^4) = \gamma_{4*}\{\nu_4\sigma'\eta_{14}, 2E\mu', \varepsilon_4\nu_{12}, (E\nu')\varepsilon_7\}.$$

Further, (2.45) and (2.51) lead to $[\gamma_6\bar{\nu}_6\nu_{14}, i_{\mathbb{R}}] = 0$ and (2.46) to $4\zeta_6 \in M_{17}(\mathbb{S}^6)$. Hence, $P_{17}(\mathbb{R}P^6) = \gamma_{6*}G_{17}(\mathbb{S}^6)$ and, by Lemma 2.12(3), $P_{23}(\mathbb{R}P^{12}) = \gamma_{12*}G_{23}(\mathbb{S}^{12})$.

(v) $(12,4), (12,6), (12,8), (12,10), (12,12)$. Recall from Sect. 1.6 that:

$$G_{n+12}(\mathbb{S}^n) = \begin{cases} \pi_{16}(\mathbb{S}^4) = \{\nu_4\sigma'\eta_{14}^2, \nu_4^4, \nu_4\mu_7, \\ \nu_4\eta_7\varepsilon_8, (E\nu')\mu_7, (E\nu')\eta_7\varepsilon_8\} & \text{for } n = 4; \\ \pi_{18}(\mathbb{S}^6) = \{[\iota_6, \iota_6]\sigma_{11}^+\} & \text{for } n = 6; \\ \pi_{20}(\mathbb{S}^8) = 0 & \text{for } n = 8; \\ \pi_{22}^{10} = \{[\iota_{10}, \nu_{10}]\} & \text{for } n = 10; \\ \pi_{24}(\mathbb{S}^{12}) = \{\theta, E\theta'\} & \text{for } n = 12. \end{cases}$$

Certainly, $P_{20}(\mathbb{R}P^8) = \pi_{20}(\mathbb{R}P^8) = 0$.

In view of (2.45) and (2.55), we get $[\gamma_4\nu_4\sigma'\eta_{14}^2, i_{\mathbb{R}}] = -2\gamma_4(E\varepsilon')\eta_{14}^2 = 0$. The relations (1.28), (2.45), and (2.49) lead to $[\gamma_4\nu_4^4, i_{\mathbb{R}}] = -(E\nu')\nu_7^3 = 0$. Again, (1.133), (2.45), and (2.49) yield $[\gamma_4\nu_4\mu_7, i_{\mathbb{R}}] = -\gamma_4(E\nu')\mu_7 \neq 0$, $[\gamma_4\nu_4\eta_7\varepsilon_8, i_{\mathbb{R}}] = -\gamma_4(E\nu')\eta_7\varepsilon_8 \neq 0$ and $[\gamma_4\nu_4(\mu_7 + \eta_7\varepsilon_8)] = \gamma_4(E\nu')(\mu_7 + \eta_7\varepsilon_8) \neq 0$. Because (2.46) leads to $[\gamma_4(E\nu')\mu_7, i_{\mathbb{R}}] = [\gamma_4(E\nu')\eta_7\varepsilon_8, i_{\mathbb{R}}] = 0$, we deduce that $P_{16}(\mathbb{R}P^4) = \gamma_{4*}\{\nu_4\sigma'\eta_{14}^2, \nu_4^4, (E\nu')\mu_7, (E\nu')\eta_7\varepsilon_8\}$.

Next, (2.46) leads to $[\gamma_{12}E\theta', i_{\mathbb{R}}] = 0$. Because $H(\theta) = \eta_{23}$ (1.66)(7), in view of Lemma 2.4(1), (1.15), (1.127), and (2.7), it holds

$$[\gamma_{12}\theta, i_{\mathbb{R}}] = \gamma_{12}E\theta' \neq 0 \tag{2.56}$$

and $P_{24}(\mathbb{R}P^{12}) = \gamma_{12*}\{E\theta'\}$. Further, Lemma 1.2(2) and Lemma 2.12(3) yield $P_{n+12}(\mathbb{R}P^n) = \gamma_{n*}\pi^n_{n+12}$ for $n = 6, 10$ as well.

(vi) $(13,4), (13,6), (13,8), (13,10), (13,12), (13,14)$. Recall from Sect. 1.6 that:

$$G_{n+13}(\mathbb{S}^n) = \begin{cases} \pi^4_{17} = \{\nu_4^2\sigma_{10}, \nu_4\eta_7\mu_8, (E\nu')\eta_7\mu_8\} & \text{for } n = 4; \\ \pi^6_{19} = \{\nu_6\sigma_9\nu_{16}\} & \text{for } n = 6, \\ \pi^8_{21} = \{\sigma_8\nu_{15}^2, \nu_8\sigma_{11}\nu_{18}\} & \text{for } n = 8, \\ \pi^{10}_{23} = \{\sigma_{10}\nu_{17}^2\} & \text{for } n = 10, \\ \pi^{12}_{25} = \{\theta\eta_{24}, (E\theta')\eta_{24}\} & \text{for } n = 12, \\ \pi^{14}_{27} = \{3[\iota_{14}, \iota_{14}]\} & \text{for } n = 14. \end{cases}$$

Certainly, $P_{n+13}(\mathbb{R}P^n) = \gamma_{n*}G_{n+13}(\mathbb{S}^n)$ for $n = 6, 10$. The relations (1.82) and (2.45) and (2.53) lead to $[\gamma_8\sigma_8\nu_{15}^2, i_{\mathbb{R}}] = \gamma_8(E\sigma')\nu_{15}^2 = \gamma_8\nu_8\sigma_{11}\nu_{18} \neq 0$. Because, in view of (2.46), it holds $[\gamma_8\nu_8\sigma_{11}\nu_{18}, i_{\mathbb{R}}] = 0$, we deduce $P_{21}(\mathbb{R}P^8) = \gamma_{8*}\{\nu_8\sigma_{11}\nu_{18}\}$.

Further, (2.46) leads to $[\gamma_{12}(E\theta')\eta_{24}, i_{\mathbb{R}}] = 0$ and, in view of (1.16), (2.45), and (2.56), it holds $[\gamma_{12}\theta\eta_{24}, i_{\mathbb{R}}] = \gamma_{12}[\iota_{12}, \eta_{12}^2] \neq 0$. Hence, $P_{25}(\mathbb{R}P^{12}) = \gamma_{12*}\{(E\theta')\eta_{24}\}$.

By (1.28), (2.45), and (2.49), we get $[\gamma_4\nu_4^2\sigma_{10}, i_{\mathbb{R}}] = -\gamma_4(E\nu')\nu_7\sigma_{10} = 0$. Again, (2.45), (2.49), and [85, Theorem 7.7] yield $[\gamma_4\nu_4\eta_7\mu_8, i_{\mathbb{R}}] = -\gamma_4(E\nu')\eta_7\mu_8 \neq 0$. Because, by (2.46), it holds $[\gamma_4(E\nu')\eta_7\mu_8, i_{\mathbb{R}}] = 0$, we derive that $P_{17}(\mathbb{R}P^4) = \gamma_{4*}\{\gamma_4\nu_4^2\sigma_{10}, \gamma_4(E\nu')\eta_7\mu_8\}$.

Finally, in view of Lemma 2.12(3), it holds $P_{27}(\mathbb{R}P^{14}) = \gamma_{14*}G_{27}(\mathbb{S}^{14})$.

Then, in light of the above and Chap. 1, the formulae (1)–(14) follow and the proof is complete. □

2.5 Some Whitehead Center Groups of Complex and Quaternionic Projective Spaces

First, we show:

Theorem 2.20. *Let $n \geq 2$. Then, $P_{k+2n+1}(\mathbb{C}P^n) = \gamma_{n*}G_{k+2n+1}(\mathbb{S}^{2n+1})$ for $0 \leq k \leq 13$ except $(k,n) = (2,2), (9,2), (10,2), (10,4), (13,4)$ and $P_7(\mathbb{C}P^2) = 0$, $P_{14}(\mathbb{C}P^2) = \gamma_{2*}\{\nu_5^3\} \cong \mathbb{Z}_2$, $P_{15}(\mathbb{C}P^2) = \gamma_{2*}\{\nu_5\sigma_8, \beta_1(5)\} \cong \mathbb{Z}_{24}$, $P_{19}(\mathbb{C}P^4) = \gamma_{4*}\{\beta_1(9)\} \cong \mathbb{Z}_3$, $P_{22}(\mathbb{C}P^4) = \gamma_{4*}\{\alpha_1(9)\beta_1(12)\} \cong \mathbb{Z}_3$, that is:*

1. $P_{2n+1}(\mathbb{C}P^n) = \begin{cases} \pi_{2n+1}(\mathbb{C}P^n) \cong \mathbb{Z} & \text{for } n = 3, \\ 2\pi_{2n+1}(\mathbb{C}P^n) \cong 2\mathbb{Z} & \text{for } n \geq 2 \text{ unless } n = 3; \end{cases}$
2. $P_{2n+k+1}(\mathbb{C}P^n) = \begin{cases} 0 & \text{for even } n, \\ \pi_{2n+k+1}(\mathbb{C}P^n) \cong \mathbb{Z}_2 & \text{for odd } n \end{cases}$ *with $k = 1, 2$;*

3. $P_{2n+4}(\mathbb{C}P^n) = \begin{cases} \pi_{2n+4}(\mathbb{C}P^n) \cong \mathbb{Z}_{24} & \textit{for } n \equiv 3 \pmod 4 \textit{ or } n = 2^i - 2 \geq 2, \\ 2\pi_{2n+4}(\mathbb{C}P^n) \cong \mathbb{Z}_{12} & \textit{for otherwise}; \end{cases}$

4. $P_{2n+k+1}(\mathbb{C}P^n) = \begin{cases} 0 & \textit{for } n \geq 3, \\ \pi_{2n+k+1}(\mathbb{C}P^n) \cong \mathbb{Z}_2 & \textit{for } n = 2 \end{cases}$ *with* $k = 4, 5$;

5. $P_{2n+7}(\mathbb{C}P^n) = \begin{cases} \pi_{2n+7}(\mathbb{C}P^n) \cong \mathbb{Z}_2 & \textit{for } n \equiv 2, 3 \pmod 4 \textit{ or } n = 2^i - 3 \geq 5, \\ 0 & \textit{for otherwise}; \end{cases}$

6. $P_{2n+8}(\mathbb{C}P^n) = \begin{cases} \pi_{12}(\mathbb{C}P^2) \cong \mathbb{Z}_{30} & \textit{for } n = 2, \\ \pi_{2n+8}(\mathbb{C}P^n) \cong \mathbb{Z}_{240} & \textit{for } n = 5 \textit{ or } n \equiv 7 \pmod 8, \\ 2\pi_{2n+8}(\mathbb{C}P^n) \cong \mathbb{Z}_{120} & \textit{for otherwise}; \end{cases}$

7. $P_{2n+9}(\mathbb{C}P^n) = \begin{cases} \pi_{13}(\mathbb{C}P^2) \cong \mathbb{Z}_2 & \textit{for } n = 2, \\ \pi_{15}(\mathbb{C}P^3) \cong (\mathbb{Z}_2)^3 & \textit{for } n = 3, \\ \gamma_{4*}\{[\iota_9, \iota_9]\} \cong \mathbb{Z}_2 & \textit{for } n = 4, \\ 0 & \textit{for } n \equiv 0 \pmod 2 \textit{ unless } n = 2, 4, \\ \pi_{2n+9}(\mathbb{C}P^n) \cong (\mathbb{Z}_2)^2 & \textit{for } n \equiv 1 \pmod 2 \textit{ unless } n = 3; \end{cases}$

8. $P_{2n+10}(\mathbb{C}P^n) = \begin{cases} \gamma_{2*}\{\nu_5^3\} \cong \mathbb{Z}_2 & \textit{for } n = 2, \\ \pi_{16}(\mathbb{C}P^3) \cong (\mathbb{Z}_2)^4 & \textit{for } n = 3, \\ \gamma_{4*}\{\sigma_9\eta_{16}^2, \nu_9^3, \eta_9\varepsilon_{10}\} \cong (\mathbb{Z}_2)^3 & \textit{for } n = 4, \\ \pi_{2n+10}(\mathbb{C}P^n) \cong (\mathbb{Z}_2)^3 & \textit{for } n \equiv 1 \pmod 2,\ n \geq 5, \\ \gamma_{n*}\{\nu_{2n+1}^3, \eta_{2n+1}\varepsilon_{2n+2}\} \cong (\mathbb{Z}_2)^2 & \textit{for } n \equiv 2 \pmod 4,\ n \geq 6 \\ & \textit{or } n = 2^i - 4,\ i \geq 4 \\ & \textit{unless } n \equiv 26 \pmod{32}, \\ \gamma_{n*}\{\nu_{2n+1}^3\} \cong \mathbb{Z}_2 & \textit{for } n \equiv 26 \pmod{32}, \\ \gamma_{n*}\{\eta_{2n+1}\varepsilon_{2n+2}\} \cong \mathbb{Z}_2 & \textit{for } n \equiv 0 \pmod 4,\ n \geq 8 \\ & \textit{unless } n = 2^i - 4; \end{cases}$

9. $P_{2n+11}(\mathbb{C}P^n) = \begin{cases} \gamma_{2*}\{\nu_5\sigma_8, \beta_1(5)\} \cong \mathbb{Z}_{24} & \textit{for } n = 2, \\ \pi_{17}(\mathbb{C}P^3) \cong \mathbb{Z}_{24} \oplus \mathbb{Z}_2 & \textit{for } n = 3, \\ \pi_{21}(\mathbb{C}P^5) \cong \mathbb{Z}_6 \oplus \mathbb{Z}_2 & \textit{for } n = 5, \\ \gamma_{n*}\{\beta_1(2n+1)\} \cong \mathbb{Z}_3 & \textit{for } n \equiv 0 \pmod 2,\ n \geq 4, \\ \pi_{2n+11}(\mathbb{C}P^n) \cong \mathbb{Z}_6 & \textit{for } n \equiv 1 \pmod 2,\ n \geq 7; \end{cases}$

10. $P_{2n+12}(\mathbb{C}P^n) = \begin{cases} \pi_{2n+12}(\mathbb{C}P^n) \cong \mathbb{Z}_{504} \oplus (\mathbb{Z}_2)^2 & \textit{for } n = 2, \\ \pi_{2n+12}(\mathbb{C}P^n) \cong \mathbb{Z}_{504} \oplus \mathbb{Z}_2 & \textit{for } n = 3, 4, \\ \pi_{2n+12}(\mathbb{C}P^n) \cong \mathbb{Z}_{504} & \textit{for } n \geq 5 \textit{ unless } n \equiv 57 \pmod{62}, \\ 2\pi_{2n+12}(\mathbb{C}P^n) \cong \mathbb{Z}_{252} & \textit{for } n \equiv 57 \pmod{62}; \end{cases}$

11. $P_{2n+13}(\mathbb{C}P^n) = \begin{cases} \pi_{17}(\mathbb{C}P^2) \cong (\mathbb{Z}_2)^3 & \textit{for } n = 2, \\ \pi_{2n+13}(\mathbb{C}P^n) = 0 & \textit{for } n = 3, 4 \textit{ or } n \geq 7, \\ \pi_{2n+13}(\mathbb{C}P^n) \cong \mathbb{Z}_2 & \textit{for } n = 5, 6; \end{cases}$

12. $P_{2n+14}(\mathbb{C}P^n) = \begin{cases} \pi_{2n+14}(\mathbb{C}P^n) \cong \mathbb{Z}_6 \oplus \mathbb{Z}_2 & \textit{for } n = 2, 5, \\ \pi_{2n+14}(\mathbb{C}P^n) \cong \mathbb{Z}_6 & \textit{for } n = 3, 6, \\ \gamma_{4*}\{\alpha_1(9)\beta_1(12)\} \cong \mathbb{Z}_3 & \textit{for } n = 4, \\ \pi_{2n+14}(\mathbb{C}P^n) \cong \mathbb{Z}_3 & \textit{for } n \geq 7. \end{cases}$

Proof. We make use of Lemma 2.4(2) to find $M_{k+2n+1,\mathbb{C}}(\mathbb{S}^{2n+1})$ for $1 \le k \le 13$ and $n = 2, 4, 6$ with $k > 2n - 1$. First, we consider $M_{k+5,\mathbb{C}}(\mathbb{S}^5)$ for $4 \le k \le 13$. By the relation (1.23), we obtain $M_{k+5,\mathbb{C}}(\mathbb{S}^5) = \pi_{k+5}(\mathbb{S}^5)$ for $4 \le k \le 6$ and $k = 12, 13$. Moreover, in view of $h_0(\sigma''') = H(\sigma''') = 4\nu_9$ (1.66)(3) and (2.7), the relations Lemma 1.2(2) and (1.23) imply $[\iota_5, \eta_5] \circ Eh_0(\sigma''') = 0$. Then, in view of (1.36), we obtain $M_{12,\mathbb{C}}(\mathbb{S}^5) = \pi_{12}(\mathbb{S}^5)$. Because $\eta_5^2\varepsilon_7 = 4\nu_5\sigma_8 \neq 0$ (1.81) and $\eta_n\mu_{n+1} \neq 0$ for $n \ge 2$ (1.62), we deduce that $M_{13,\mathbb{C}}(\mathbb{S}^5) = 0$, $M_{14,\mathbb{C}}(\mathbb{S}^5) = \{\nu_5^3\}$, and $M_{15,\mathbb{C}}(\mathbb{S}^5) = \{\nu_5\sigma_8, \beta_1(5)\}$. The relation (1.140) yields $M_{16,\mathbb{C}}(\mathbb{S}^5) = \pi_{16}(\mathbb{S}^5)$.

To find $M_{k+9,\mathbb{C}}(\mathbb{S}^9)$ for $8 \le k \le 13$, we recall [85, Theorems 7.1–7.4 and 7.6–7.7]:

$$\pi_{k+9}(\mathbb{S}^9) = \begin{cases} \{\sigma_9\eta_{16}, \bar{\nu}_9, \varepsilon_9\} & \text{for } k = 8, \\ \{\sigma_9\eta_{16}^2, \nu_9^3, \mu_9, \eta_9\varepsilon_{10}\} & \text{for } k = 9, \\ \{\sigma_9\nu_{16}, \eta_9\mu_{10}, \beta_1(9)\} & \text{for } k = 10, \\ \{\zeta_9 + \alpha_3'(9) + \alpha_{1,7}(9), \bar{\nu}_9\nu_{17}\} & \text{for } k = 11, \\ 0 & \text{for } k = 12, \\ \{\sigma_9\nu_{16}^2, \alpha_1(9)\beta_1(12)\} & \text{for } k = 13. \end{cases}$$

In view of [85, Theorem 7.2], (1.69)(1), and (1.72), we get $\eta_9\sigma_{10}\eta_{17} = \nu_9^3 + \eta_9\varepsilon_{10} \neq 0$, $\eta_9\bar{\nu}_{10} = \nu_9^3 \neq 0$, $\eta_9\varepsilon_{10} \neq 0$, and $\eta_9(E[\iota_9, \iota_9]) = 0$. Consequently, by means of (1.70), it holds $M_{17,\mathbb{C}}(\mathbb{S}^9) = \{\sigma_9\eta_{16} + \bar{\nu}_9 + \varepsilon_9\} = \{[\iota_9, \iota_9]\}$.

By (1.82) and (1.81), it holds $\eta_9^2\varepsilon_{11} = 4\nu_9\sigma_{12} = 0$. Then, in view of (1.18) and (1.69)(1), we get $\eta_9\sigma_{10}\eta_{17}^2 = \eta_9^3\sigma_{12} = 4\nu_9\sigma_{12} = 0$. Next, $\eta_9\nu_{10}^3 = 0$ (1.23) and $\eta_9\mu_{10} \neq 0$ (1.62). Hence, $M_{18,\mathbb{C}}(\mathbb{S}^9) = \{\sigma_9\eta_{16}^2, \nu_9^3, \eta_9\varepsilon_{10}\}$. In view of (1.125) and (1.64)(2), it holds $M_{19,\mathbb{C}}(\mathbb{S}^9) = \{\beta_1(9)\}$.

By $\eta_9\zeta_{10} = 0$ (1.140) and $\bar{\nu}_{10}\nu_{18} = 0$ (1.129), we get $M_{20,\mathbb{C}}(\mathbb{S}^9) = \pi_{20}(\mathbb{S}^9)$. Certainly, $\pi_{21}(\mathbb{S}^9) = 0$ implies $M_{21,\mathbb{C}}(\mathbb{S}^9) = 0$. The relation (1.109)(2) yields $M_{22,\mathbb{C}}(\mathbb{S}^9) = \{\alpha_1(9)\beta_1(12)\}$.

In virtue of (1.128), we get that $M_{25,\mathbb{C}}(\mathbb{S}^{13}) = \pi_{25}(\mathbb{S}^{13})$ and $M_{26,\mathbb{C}}(\mathbb{S}^{13}) = \pi_{26}(\mathbb{S}^{13})$.

Notice that for $k \le 4n - 1$, we have

$$E^{-1}(\text{Ker } \eta_{4n+1*}) = \begin{cases} 2\pi_{4n+k+1}(\mathbb{S}^{4n+1}) & \text{for } k = 0, 7, \\ \{\nu_{4n+1}^3, \eta_{4n+1}\varepsilon_{4n+2}\} & \text{for } k = 9, \\ 0 & \text{for } k = 1, 2, 8, 10, 12, \\ \pi_{4n+k+1}(\mathbb{S}^{4n+1}) & \text{for } 3 \le k \le 6,\ k = 11, 13. \end{cases}$$

This, (1.2)–(1.50), Proposition 2.16(1);(3) and Chap. 1 lead to the assertion and the proof is complete. □

We recall:

$$\pi_k(\mathbb{S}^3) = \begin{cases} \{\nu'\eta_6^{k-6}\} \cong \mathbb{Z}_2 & \text{for } k = 7, 8, \\ \{\varepsilon_3\} \cong \mathbb{Z}_2 & \text{for } k = 11, \\ \{\eta_3\varepsilon_4, \mu_3\} \cong (\mathbb{Z}_2)^2 & \text{for } k = 12. \end{cases}$$

We also need:

$$\pi_k(\mathbb{S}^3) = \begin{cases} \{\nu'\mu_6, \nu'\eta_6\varepsilon_7\} \cong (\mathbb{Z}_2)^2 & \text{for } k = 15, \\ \{\nu'\eta_6\mu_7, \alpha_1(3)\beta_1(6)\} \cong \mathbb{Z}_6 & \text{for } k = 16, \\ \{\varepsilon_3\nu_{11}^2, \alpha_1(3)\alpha_3'(6), \alpha_{1,5}(3)\alpha_{1,5}(10)\} \cong \mathbb{Z}_{30} & \text{for } k = 17, \\ \{\bar{\varepsilon}_3, \alpha_4(3), \alpha_{2,5}(3)\} \cong \mathbb{Z}_{30} & \text{for } k = 18. \end{cases}$$

Notice that (2.10) and Lemma 2.6(3) lead to:

$$\nu' \circ \pi_k^6 \subseteq Q_k^3. \tag{2.57}$$

Further, we show:

Lemma 2.21. 1. $Q_k(\mathbb{S}^3) = 0$ *for* $k = 11, 12$;
2. $Q_k(\mathbb{S}^3) = \pi_k(\mathbb{S}^3) \cong \mathbb{Z}_2$ *for* $k = 7, 8$ *and* $Q_9(\mathbb{S}^3) = \pi_9(\mathbb{S}^3) \cong \mathbb{Z}_3$;
3. $Q_{10}(\mathbb{S}^3) = 3\pi_{10}(\mathbb{S}^3) \cong \mathbb{Z}_5$ *and* $Q_{13}(\mathbb{S}^3) = \{\varepsilon', \alpha_1(3)\alpha_2(6)\} \cong \mathbb{Z}_{12}$;
4. $Q_{14}(\mathbb{S}^3) = \pi_{14}(\mathbb{S}^3) \cong \mathbb{Z}_{84} \oplus (\mathbb{Z}_2)^2$;
5. $Q_{15}(\mathbb{S}^3) = \pi_{15}(\mathbb{S}^3) \cong (\mathbb{Z}_2)^2$, $Q_{16}(\mathbb{S}^3) = \pi_{16}^3 = \{\nu'\eta_6\mu_7\} \cong \mathbb{Z}_2$, $Q_{17}(\mathbb{S}^3) = \pi_{17}(\mathbb{S}^3) \cong \mathbb{Z}_{30}$ *and* $Q_{18}(\mathbb{S}^3) = 6\pi_{18}(\mathbb{S}^3) \cong \mathbb{Z}_5$.

Proof. It is known that $\nu' E^3\alpha \neq 0$ for $\alpha = \varepsilon_3, \eta_3\varepsilon_4, \mu_3$, and $\eta_3\mu_4$. This, (2.10), and (2.11) lead to (1) and the fact that $Q_{13}(\mathbb{S}^3) \not\ni \eta_3\mu_4$.

Notice that $\pi_{k+6}^3 = \nu' \circ \pi_{k+6}^6$ for $k = 1, 2$ and $\alpha_1(3)\alpha_1(6)\alpha_1(9) = 0$ (1.8). Hence, (2.10), (2.11), and (2.57) imply (2).

Because $\alpha_1(3)\alpha_2(6) \neq 0$ and $\alpha_1(3)\alpha_{1,5}(6) = 0$, the relations (2.10) and (2.11) lead to $Q_{10}(\mathbb{S}^3) = \{\alpha_{1,5}(3)\} \cong \mathbb{Z}_5$.

In view of (1.9), we get $\alpha_1(3)\alpha_1(6)\alpha_2(9) = -3\alpha_1(3)\beta_1(6) = 0$. Then, (2.10), (2.11), (1.124), and (2.57) yield (3).

Next, $\pi_{14}^3 = \{\mu', \varepsilon_3\nu_{11}, \nu'\varepsilon_6\} \cong \mathbb{Z}_4 \oplus (\mathbb{Z}_2)^2$. In view of (1.18), (1.64), and (1.140), we obtain $\nu' E^3\mu' = \eta_3^3\zeta_6 = 0$. By means of (1.122), the relation (2.57) implies $Q_{14}^3 = \pi_{14}^3$. Then, (2.10), (2.11), and $\alpha_1(3)\alpha_{1,7}(6) = 0$, $\alpha_1(3)\alpha_3(6) = 3\alpha_1(3)\alpha_3'(6) = 0$ (1.10) lead to (4).

Observe that the relations (2.10), (2.11), and (1.9) lead to $\alpha_1(3)\beta_1(6) \notin Q_{16}(\mathbb{S}^3)$. Because $\nu' \circ \pi_k^6 = \pi_k^3$ for $k = 15, 16$, the relation (2.57) leads to the first and second one of (5).

Because $\pi_{17}^3 = \nu' \circ \pi_{17}^6$ (1.122), the relation (2.57) yields $Q_{17}^3 = \pi_{17}^3$. Then, (2.10), (2.11), and (2.37) lead to the third one of (5).

We know that $\alpha_1(3)\alpha_4(6) \neq 0$ Proposition 2.11(3). This, $\alpha_1(3)\alpha_{2,5}(6) = 0$, (2.10), (2.11), and (2.21) lead to the last of (5), and the proof is complete. □

Notice that by (1.1), Propositions 1.6, 1.5, and [85, Proposition 2.5],

$$[\iota_n, \pi_k^n] = J(\Delta\pi_k^n) = P(E^{n+1}\pi_k^n).$$

We show:

Proposition 2.22. *Let $4 \leq k \leq 22$. Then, $\nu_6 E^6 \pi_k^3 \subseteq P\pi_{k+8}^{13}$, $\nu_7 E^7 \pi_k(\mathbb{S}^3) = 0$ for $4 \leq k \leq 21$, and $\nu_k E^k \pi_{22}(\mathbb{S}^3) \cong \mathbb{Z}_3$ for $5 \leq k \leq 9$.*

Proof. We know $\nu_5 \circ E^5\nu' = 2\nu_5^2 = 0$ (1.25). This implies $\nu_6 E^6 \pi_k^3 = 0$ for $4 \leq k \leq 10$.

The relations (2.30), (1.126), (1.39), and (2.34) imply $\nu_6 E^6 \pi_{11}^3 = P\pi_{19}^{13}$, $\nu_6 E^6 \pi_{12}^3 = 8P\pi_{20}^{13}$, and $\nu_6 E^6 \mu' = 0$. By $\pi_{18}^5 \cong (\mathbb{Z}_2)^2$, we obtain $\nu_5 E^5 \varepsilon' \in \{\nu_5, 2\nu_8, \nu_{11}\} \circ 2\nu_{15} \subseteq 2\pi_{18}^5 = 0$. Consequently, $\nu_6 E^6(\nu'\varepsilon_6) = 0$. The relation (1.126) implies $\nu_6\varepsilon_9\nu_{17} = 2\bar{\nu}_6 \circ \nu_{14}^2 = 0$. In view of (1.18), (1.122), and the relation (2.20), we get

$$\pi_k^3 = \nu' \circ \pi_k^6 \text{ for } 15 \leq k \leq 17.$$

Therefore, we obtain $\nu_6 E^6 \pi_k^3 = 0$ for $15 \leq k \leq 17$.

Since $\pi_{18}^3 = \{\bar{\varepsilon}_3\} \cong \mathbb{Z}_2$, (1.108) implies $\nu_6 E^6 \pi_{18}^3 = 0$.

Further $\pi_{19}^3 = \{\eta_3\bar{\varepsilon}_4, \mu_3\sigma_{12}\} \cong (\mathbb{Z}_2)^2$, $\nu_6\eta_9\bar{\varepsilon}_{10} = 0$ (1.23), and $\nu_6\mu_9\sigma_{18} = 8P(\sigma_{13}^2)$ (1.39) lead to $\nu_6 E^6 \pi_{18}^3 = 8\{P(\sigma_{13}^2)\}$.

Next, we recall

$$\pi_{20}^3 = \{\bar{\varepsilon}', \bar{\mu}_3, \eta_3\mu_4\sigma_{13}\} \cong \mathbb{Z}_4 \oplus (\mathbb{Z}_2)^2.$$

Then, the relations (2.32) and (2.22) imply $\nu_6 E^6 \pi_{20}^3 = 16\{P(\rho_{13})\}$.

From the fact that $\pi_{21}^3 = \{\mu'\sigma_{14}, \nu'\bar{\varepsilon}_6, \eta_3\bar{\mu}_4\}$ and (2.34), we obtain $\nu_6 E^6 \pi_{21}^3 = 0$. The relation (2.35) implies $\nu_6 E^6 \pi_{22}^3 = 0$.

Next, we examine the case for odd primes. Those cases appear for $k = 6, 9, 10, 13, 14$ and $16 \leq k \leq 22$. By (1.8), the assertion

$$\alpha_1(7)E^7\pi_k(\mathbb{S}^3; 3) = 0 \tag{2.58}$$

holds for $k = 6, 9, 13, 16$.

By (1.9) and (1.10), (2.58) holds for $k = 10, 14$.

By (1.8) and Proposition 2.11(1), (2.58) holds for $k = 17$.

For $k = 18$, we have $\alpha_1(7)E^7\pi_{18}(\mathbb{S}^3; 3) = \{\alpha_1(7)\alpha_4(10)\} \subseteq E^2\pi_{23}(\mathbb{S}^5; 3) = 0$ [85, p. 185].

By [85, Theorem 13.10(iv);(v)], $E^2\pi_{l+19}(\mathbb{S}^3; 3) = 0$ for $l = 0, 1$, and hence, (2.58) holds for $k = 19, 20$.

By [85, p. 185], $E^2\pi_{21}(\mathbb{S}^3; 3) = 0$, and hence, (2.58) for $k = 21$ holds.

Finally, for $k = 22$, by Proposition 2.11(4) and [87, p. 56], $\alpha_1(9)\alpha_5(12) \neq 0$. This leads to the last result and completes the proof. □

We propose:

Problem 2.23. If $\nu' E^3\beta = 0$ for $\beta \in \pi_{k-1}^3$ then $\nu_4 \wedge \beta = 0$?

Lemma 2.8(1) and Proposition 2.22 yield:

Corollary 2.24. *$L''_{k-1,n}(\mathbb{S}^3) = \pi_{k-1}(\mathbb{S}^3)$ for $n \geq 2$ and $5 \leq k \leq 22$.*

Writing R_k^n for the subgroup of $\pi_k(SO(n))$ consisting of its free part and $\pi_k(SO(n);2)$, we get from $(\mathcal{SF}_k^n)$ the exact sequence

$$(\mathcal{R}_k^n)\quad \cdots \longrightarrow \pi_{k+1}^n \xrightarrow{\Delta} R_k^n \xrightarrow{i_*} R_k^{n+1} \xrightarrow{p_*} \cdots .$$

Now, we show:

Theorem 2.25. *$P_{4n+3+k}(\mathbb{H}P^n) = P'_{4n+3+k}(\mathbb{H}P^n) \oplus P''_{4n+3+k}(\mathbb{H}P^n)$ for $0 \le k \le 10$, where:*

1. $P'_{4n+3}(\mathbb{H}P^n) = [[\frac{24}{(24,n+1)}, 2]]\gamma_{n*}\pi_{4n+3}(\mathbb{S}^{4n+3})$ *for $n \ge 2$;*
2. $P'_{4n+3+k}(\mathbb{H}P^n) = \gamma_{n*}\pi_{4n+k+3}(\mathbb{S}^{4n+3})$ *for $k = 1, 2, 4, 5, 8, 9, 10$;*
3. $P'_{4n+6}(\mathbb{H}P^n) = \frac{3+(-1)^n}{2}\gamma_{n*}\pi_{4n+6}(\mathbb{S}^{4n+3})$;
4. $P'_{4n+9}(\mathbb{H}P^n) = \frac{1-(-1)^n}{2}\gamma_{n*}\pi_{4n+9}(\mathbb{S}^{4n+3})$;
5. $P'_{4n+10}(\mathbb{H}P^n) = \begin{cases} 2\gamma_{n*}\pi_{4n+10}(\mathbb{S}^{4n+3}), & \text{for } n \equiv 0, 1, 2 \pmod 4; \\ \gamma_{n*}\pi_{4n+10}(\mathbb{S}^{4n+3}), & \text{for } n \equiv 3 \pmod 4 \text{ or } n = 2. \end{cases}$

Proof. First, recall that in view of Proposition 2.10, it holds $P'_k(\mathbb{H}P^n) = \gamma_{n*}(G_k(\mathbb{S}^{4n+3}) \cap M_k(\mathbb{S}^{4n+3}))$. Because the groups $G_{4n+k+3}(\mathbb{S}^{4n+3})$ for $0 \le k \le 10$ have been determined in Sects. 1.2 and 1.6, Proposition 2.10 and Lemma 2.12(2)–(iii) yield

$$P'_{4n+k+3}(\mathbb{H}P^n) = \begin{cases} [[\frac{24}{(24,n+1)}, 2]]\gamma_{n*}\pi_{4n+3}(\mathbb{S}^{4n+3}) & \text{for } k = 0; \\ \gamma_{n*}\pi_{4n+k+3}(\mathbb{S}^{4n+3}) & \text{for } k = 1, 2, 4, 5; \\ \frac{3+(-1)^n}{2}\gamma_{n*}\pi_{4n+6}(\mathbb{S}^{4n+3}) & \text{for } k = 3; \\ \frac{3+(-1)^n}{2}\gamma_{n*}\pi_{4n+6}(\mathbb{S}^{4n+3}) & \text{for } k = 6; \\ 2\gamma_{n*}\pi_{4n+10}(\mathbb{S}^{4n+3}) & \text{for } k = 7 \text{ with } n \equiv 0, 1, 2 \pmod 4; \\ \gamma_{n*}\pi_{4n+10}(\mathbb{S}^{4n+3}) & \text{for } k = 7 \text{ with } n \equiv 3 \pmod 4 \text{ or } n = 2; \\ \gamma_{n*}\pi_{4n+k+3}(\mathbb{S}^{4n+3}) & \text{for } k = 8, 9, 10 \end{cases}$$

and (1)–(5) follow.

Next, notice that by means of Lemma 2.8(1)–(2), Proposition 2.10, and Proposition 2.17(1), we get

$$P_k(\mathbb{H}P^{2n+1}) = P'_k(\mathbb{H}P^{2n+1}) \oplus P''_k(\mathbb{H}P^{2n+1}).$$

Now, we show that

$$P_{8n+3+k}(\mathbb{H}P^{2n}) = P'_{8n+3+k}(\mathbb{H}P^{2n}) \oplus P''_{8n+3+k}(\mathbb{H}P^{2n})$$

for $0 \le k \le 10$. By Lemma 2.8(2)–(3) and Proposition 2.17(2)(i)–(ii), we see that it holds for p-primary components with an odd prime p. To show that for the 2-primary component, we check the condition (2.43) in Proposition 2.17(2)-(iii). Let n be even and we work in the 2-primary component. The assertion is a direct consequence of the fact that $[\iota_{4n+3}, \pi_{4n+3+k}^{4n+3}] = 0$ for $k = 1, 2, 8, 9, 10$ [20, Proposition 1.3].

For $k = 0, 3, 6$, the condition (2.43) holds because $2[\iota_{4n+3}, \iota_{4n+3}] = 2[\iota_{4n+3}, \nu_{4n+3}] = 2[\iota_{4n+3}, \nu^2_{4n+3}] = 0$.

We know $[\iota_{4n+3}, \sigma_{4n+3}] = 0$ if and only if $n = 2$ or $n \equiv 3 \pmod 4$ (1.50). So, for the case $n \neq 2$ and $n \equiv 0, 2 \pmod 4$, suppose that there exists $\beta \in \pi^3_{4n+9}$, such that $[\iota_{4n+3}, \sigma_{4n+3}] = \nu_{4n+3}E^{4n+3}\beta$. By [10] and [30],

$$R^{4n+k}_{4n+9} \cong \begin{cases} \mathbb{Z}_2 \oplus \mathbb{Z}_8 \oplus \mathbb{Z}_2 & \text{for } k = 1; \\ \mathbb{Z}_2 \oplus \mathbb{Z}_4 & \text{for } k = 2; \\ \mathbb{Z}_2 \oplus \mathbb{Z}_2 & \text{for } k = 3; \\ \mathbb{Z}_2 & \text{for } k = 4. \end{cases}$$

Here, the first direct summand $\mathbb{Z}_2$ corresponds to the Bott result: $R^{\infty}_{4n+9} \cong \mathbb{Z}_2$. Since $n \equiv 0, 2 \pmod 4$, in view of [2, Theorem 1.3], the map $J : R^{\infty}_{4n+9} \to \pi^S_{4n+9}$ is a monomorphism. But, by means of (1.50), it holds $[\iota_{4n+3}, \sigma_{4n+3}] \neq 0$ and this element does not correspond to the Bott one. Hence, the map $J : R^{4n+3}_{4n+9} \to \pi^{4n+3}_{4n+9}$ is also a monomorphism and, by means of (2.44), it holds $\Delta(\sigma_{4n+3}) = [\iota_3]_{4n+3}\beta$.

In the exact sequence $(\mathcal{R}^{4n+k-1}_{4n+9})$ $(k = 2, 3)$:

$$R^{4n+k-1}_{4n+9} \xrightarrow{i_*} R^{4n+k}_{4n+9} \xrightarrow{p_*} \pi^{4n+k-1}_{4n+9},$$

we know $\sharp[\iota_{4n+2}, \sigma_{4n+2}] = 16$ (1.50) and $[\iota_{4n+1}, \eta_{4n+1}\sigma_{4n+2}] \neq 0$, $[\iota_{4n+1}, \bar{\nu}_{4n+1}] \neq 0$, and $[\iota_{4n+1}, \varepsilon_{4n+1}] \neq 0$ [20, Lemma 4.3]. So, $i_* : R^{4n+k-1}_{4n+9} \to R^{4n+k}_{4n+9}$ are epimorphisms for $k = 2, 3$, respectively. This shows that $[\iota_3]_{4n+1}\beta$ generates the direct summand $\mathbb{Z}_8$ in R^{4n+1}_{4n+9} and contradicts the fact that $4\beta = 0$ [32, Corollary (1.22)]. Hence, we have the assertion for $k = 7$ and this completes the proof. □

For n even, since $[\iota_{8n+3}, \sigma^2_{8n+3}] \neq 0$ [66, Table 1], the assertion of Theorem 2.25 for $k = 7$, $n \equiv 0 \pmod 4$ is obtained from Corollary 2.18(2).

Since ζ_n is cyclic except for $n \not\equiv 115 \pmod{128}$ [20, p. 426], we obtain:

Remark 2.26. The assertion of Theorem 2.25 for $k = 11$ holds and $P'_{4n+14}(\mathbb{H}P^n) = \gamma_{n*}\pi_{4n+14}(\mathbb{S}^{4n+3})$, for $n \not\equiv 115 \pmod{128}$.

Since the nontrivial Whitehead product cannot be desuspended too much, we add:

Conjecture 2.27. The assertion $P_{4n+3+k}(\mathbb{H}P^n) = P'_{4n+3+k}(\mathbb{H}P^n) \oplus P''_{4n+3+k}(\mathbb{H}P^n)$ holds all $k \geq 1$.

Now, we show:

Lemma 2.28. 1. $\mathrm{Ker}\{E^4 : \pi_{k-1}(\mathbb{S}^3) \to \pi_{k+3}(\mathbb{S}^7)\} \subseteq (i_{\mathbb{H}}E)^{-1}_* P_k(\mathbb{H}P^n)$ *for* $k, n \geq 1$. *In particular,* $i_{\mathbb{H}}E(\nu'\alpha) \in P''_k(\mathbb{H}P^n)$ *provided that* $\sharp\alpha = 2$ *for* $\alpha \in \pi^7_{k-1}$.
2. *If* $i_{\mathbb{H}}E(\nu'^{+})(E^4\alpha) \neq 0$ *for* $\alpha \in \pi_{n-1}(\mathbb{S}^3)$ *then* $i_{\mathbb{H}}(E\alpha) \notin P_n(\mathbb{H}P^2)$.

Proof. (1) is obtained from the formula

$$[i_{\mathbb{H}}E\alpha,\beta] = [i_{\mathbb{H}},\beta]\circ E^m\alpha \text{ for } \beta\in\pi_m(\mathbb{H}P^n) \text{ Lemma 1.2(2).} \tag{2.59}$$

In view of (2.14), it holds $\pm[i_{\mathbb{H}}(E\alpha), i_{\mathbb{H}}] = i_{\mathbb{H}}(E\nu'^{+})(E^4\alpha)$ for $\alpha\in\pi_{n-1}(\mathbb{S}^3)$ and (2) follows. □

Let $n\geq 2$. Then, immediately, we get:

Example 2.29.

$$P_k(\mathbb{H}\mathrm{P}^n) \ni \begin{cases} i_{\mathbb{H}}E(\nu'\eta_6) & \text{for } k=8,\\ i_{\mathbb{H}}E(\nu'\eta_6^2) & \text{for } k=9,\\ i_{\mathbb{H}}E(\alpha_1(3)\alpha_1(6)) & \text{for } k=10,\\ i_{\mathbb{H}}E(\nu'\varepsilon_6) & \text{for } k=15,\\ i_{\mathbb{H}}E(\nu'\mu_6), i_{\mathbb{H}}E(\nu'\eta_6\varepsilon_7) & \text{for } k=16,\\ i_{\mathbb{H}}E(\nu'\eta_6\mu_7) & \text{for } k=17,\\ i_{\mathbb{H}}E(\varepsilon_3\nu_{11}^2) & \text{for } k=18.\end{cases}$$

By Proposition 2.17, Lemma 2.21, Theorem 2.25, and the above, we obtain:

Proposition 2.30. 1. *Let $n\geq 2$. Then:*

$$P_k(\mathbb{H}P^n) = \begin{cases} 0 & \textit{for } k=5,6,\\ i_{\mathbb{H}*}E\{\nu'\}\cong\mathbb{Z}_4 & \textit{for } k=7,\\ i_{\mathbb{H}*}E\{\nu'\eta_6\}\cong\mathbb{Z}_2 & \textit{for } k=8,\\ i_{\mathbb{H}*}E\{\nu'\eta_6^2\}\cong\mathbb{Z}_2 & \textit{for } k=9,\\ i_{\mathbb{H}*}E\{\alpha_1(3)\alpha_1(6)\}\cong\mathbb{Z}_3 & \textit{for } k=10;\end{cases}$$

2. $$P_k(\mathbb{H}P^2) = \begin{cases} \{8\gamma_2\}\oplus i_{\mathbb{H}*}E\{\alpha_{1,5}(3)\}\cong 8\mathbb{Z}\oplus\mathbb{Z}_5 & \textit{for } k=11,\\ \{\gamma_2\eta_{11}\}\cong\mathbb{Z}_2 & \textit{for } k=12,\\ \{\gamma_2\eta_{11}^2\}\cong\mathbb{Z}_2 & \textit{for } k=13;\end{cases}$$

3. $$P_k(\mathbb{H}P^3) = \begin{cases} \{6\gamma_3\}\oplus i_{\mathbb{H}*}E\{\mu'^{+},\varepsilon_3\nu_{11},\nu'\varepsilon_6\} & \\ \cong 6\mathbb{Z}\oplus\mathbb{Z}_{84}\oplus(\mathbb{Z}_2)^2 & \textit{for } k=15,\\ \{\gamma_3\eta_{15}\}\oplus i_{\mathbb{H}*}E\{\nu'\mu_6,\nu'\eta_6\varepsilon_7\}\cong(\mathbb{Z}_2)^3 & \textit{for } k=16,\\ \{\gamma_3\eta_{15}^2\}\oplus i_{\mathbb{H}*}E\{\nu'\eta_6\mu_7\}\cong(\mathbb{Z}_2)^2 & \textit{for } k=17,\\ \{\gamma_3\nu_{15},\gamma_3\alpha_1(15)\}\oplus i_{\mathbb{H}*}E\{\varepsilon_3\nu_{11}^2, & \\ \alpha_1(3)\alpha_3'(6),\alpha_{1,5}(3)\alpha_{1,5}(10)\}\cong\mathbb{Z}_{24}\oplus\mathbb{Z}_{30} & \textit{for } k=18,\end{cases}$$ *where*
$\mu'^{+} = \mu' + \alpha_3(3) + \alpha_{1,7}(3)$;

4. *Let $n\geq 3$. Then:*

$$P_k(\mathbb{H}P^n) = \begin{cases} i_{\mathbb{H}*}E\{\alpha_{1,5}(3)\}\cong\mathbb{Z}_5 & \textit{for } k=11,\\ 0 & \textit{for } k=12,13,\\ i_{\mathbb{H}*}E\{\varepsilon',\alpha_1(3)\alpha_2(6)\}\cong\mathbb{Z}_{12} & \textit{for } k=14.\end{cases}$$

2.6 Gottlieb Groups of Real Projective Spaces

Hereafter, we set

$$G'_k(\mathbb{F}P^n) = G_k(\mathbb{F}P^n) \cap \gamma_{n*}\pi_k(\mathbb{S}^{d(n+1)-1}).$$

Notice that

$$G'_k(\mathbb{F}P^n) = G_k(\mathbb{F}P^n) \ \text{ if } \mathbb{F} = \mathbb{R}, \mathbb{C} \text{ and } k \geq d+1. \tag{2.60}$$

First of all, we show:

Proposition 2.31. $G'_k(\mathbb{F}P^n) \subseteq \gamma_{n*}G_k(\mathbb{S}^{d(n+1)-1})$.

Proof. The real case is just [22, Theorem 6-2]. For any element $\gamma_n\alpha \in G'_k(\mathbb{F}P^n)$, we have $0 = [\gamma_n\alpha, \gamma_n] = \gamma_n \circ [\alpha, \iota_{d(n+1)-1}]$. Since $\gamma_{n*} : \pi_{k+d(n+1)-2}(\mathbb{S}^{d(n+1)-1}) \to \pi_{k+d(n+1)-2}(\mathbb{F}P^n)$ is a monomorphism, $\alpha \in P_k(\mathbb{S}^{d(n+1)-1}) = G_k(\mathbb{S}^{d(n+1)-1})$. □

Let K be a closed subgroup of a Lie group H and write H/K for the left coset. We recall [39, Theorem II.5] and [81, Example 2.2] which are directly obtained by Lemma 2.1 and the fact that the usual pairing $H \times H/K \to H/K$ is an associated map for the projection $p : H \to H/K$:

Lemma 2.32. *Let K be a closed subgroup of a Lie group H and H/K the left coset. Then, the projection $p : H \to H/K$ is cyclic and $p_*\pi_n(H) \subseteq G_n(H/K)$ for $n \geq 1$.*

Write $i'_{n,\mathbb{F}} : O_{\mathbb{F}}(n-1) \times O_{\mathbb{F}}(1) \hookrightarrow O_{\mathbb{F}}(n)$ for the inclusion map and $p'_{n,\mathbb{F}} : O_{\mathbb{F}}(n) \to \mathbb{F}P^{n-1}$ for the quotient one. Now, we consider the exact sequence induced by the fibration $O_{\mathbb{F}}(n+1) \xrightarrow{O_{\mathbb{F}}(n)\times O_{\mathbb{F}}(1)} \mathbb{F}P^n$:

$$(\mathcal{FP}^n_k) \quad \cdots \longrightarrow \pi_k(\mathbb{F}P^n) \xrightarrow{\Delta'_{\mathbb{F}}} \pi_{k-1}(O_{\mathbb{F}}(n) \times O_{\mathbb{F}}(1)) \xrightarrow{i'_*} \pi_{k-1}(O_{\mathbb{F}}(n+1)) \xrightarrow{p'_*} \cdots,$$

where $i' = i'_{n+1,\mathbb{F}}$ and $p' = p'_{n+1,\mathbb{F}}$. Then, by Lemma 2.32, we have

$$\operatorname{Ker} \Delta'_{\mathbb{F}} = p'_*\pi_k(O_{\mathbb{F}}(n+1)) \subseteq G_k(\mathbb{K}P^n). \tag{2.61}$$

Next, we consider the natural map from $(\mathcal{SF}^n_k)$ to $(\mathcal{FP}^n_k)$:

$$\begin{array}{ccccc}
\pi_k(SO_{\mathbb{F}}(n+1)) & \xrightarrow{p_*} & \pi_k(\mathbb{S}^{d(n+1)-1}) & \xrightarrow{\Delta_{\mathbb{F}}} & \pi_{k-1}(SO_{\mathbb{F}}(n)) \\
\| & & \downarrow{\scriptstyle \gamma_{n*}} & & \downarrow{\scriptstyle \cap} \\
\pi_k(O_{\mathbb{F}}(n+1)) & \xrightarrow{p'_*} & \pi_k(\mathbb{F}P^n) & \xrightarrow{\Delta'_{\mathbb{F}}} & \pi_{k-1}(O_{\mathbb{F}}(n) \times O_{\mathbb{F}}(1)).
\end{array} \tag{2.62}$$

We show a key lemma determining $G_k(\mathbb{F}P^n)$:

Lemma 2.33. 1. $\mathrm{Ker}\{\Delta_{\mathbb{F}} : \pi_k(\mathbb{S}^{d(n+1)-1}) \to \pi_{k-1}(SO_{\mathbb{F}}(n))\} \subseteq \gamma_{n*}^{-1} G_k(\mathbb{F}P^n)$. *In particular, it holds* $\gamma_n \pi_k(\mathbb{S}^{d(n+1)-1}; p) \subseteq G_k(\mathbb{F}P^n)$ *provided* $\pi_{k-1}(SO_{\mathbb{F}}(n); p) = 0$ *for a prime* p.

2. *Let* $k \geq d+1$. *If* $E^{d-1} \circ J_{\mathbb{F}} \mid_{\Delta_{\mathbb{F}}(\pi_k(\mathbb{S}^{d(n+1)-1}))}: \Delta_{\mathbb{F}}\pi_k(\mathbb{S}^{d(n+1)-1}) \to \pi_{k+d(n+1)-2}(\mathbb{S}^{d(n+1)-1})$ *is a monomorphism then* $\gamma_{n*} G_k(\mathbb{S}^{d(n+1)-1}) \subseteq G_k(\mathbb{F}P^n)$. *In particular, under the assumption,*

$$G_k(\mathbb{F}P^n) = \gamma_{n*} G_k(\mathbb{S}^{d(n+1)-1}) \textit{ for } \mathbb{F} = \mathbb{R}, \mathbb{C} \textit{ and } G'_k(\mathbb{H}P^n) = \gamma_{n*} G_k(\mathbb{S}^{4n+3}).$$

Proof. By the commutativity of the right square of the diagram (2.62), Ker $\Delta_{\mathbb{F}} = \gamma_{n*}^{-1}(\mathrm{Ker}\ \Delta'_{\mathbb{F}})$. Hence, (2.61) implies (1).

(2) By the assumption, (1.1), (1.49), and Proposition 1.5 we get $G_k(\mathbb{S}^{d(n+1)-1}) = \mathrm{Ker}(J \circ \Delta) = \mathrm{Ker}\ \Delta_{\mathbb{F}}$. So, (1) leads to $\gamma_{n*} G_k(\mathbb{S}^{d(n+1)-1}) \subseteq G_k(\mathbb{F}P^n)$ and, consequently, we obtain $\gamma_{n*} G_k(\mathbb{S}^{d(n+1)-1}) \subseteq G'_k(\mathbb{F}P^n)$. On the other hand, Proposition 2.31 yields $G'_k(\mathbb{F}P^n) \subseteq \gamma_{n*} G_k(\mathbb{S}^{d(n+1)-1})$. This and (2.60) complete the proof. □

By Proposition 2.31, Lemma 2.33(1), and the fact that

$$\pi_{k-1}(SO(2)) = 0 \text{ for } k \geq 3, \tag{2.63}$$

we obtain:

Corollary 2.34. $\mathrm{Ker}\{\Delta : \pi_k(\mathbb{S}^n) \to \pi_{k-1}(SO(n))\} \subseteq \gamma_{n*}^{-1} G_k(\mathbb{R}P^n) \subseteq G_k(\mathbb{S}^n)$.
In particular:

1. $G_k(\mathbb{R}P^n) = \pi_k(\mathbb{R}P^n)$ *provided that* $\Delta\pi_k(\mathbb{S}^n) = 0$ *for* $k \geq 2$;
2. $G_k(\mathbb{S}^2) = \pi_k(\mathbb{S}^2)$ *and* $G_k(\mathbb{R}P^2) = \pi_k(\mathbb{R}P^2)$ *for* $k \geq 3$.

By [21], we know:

Theorem 2.35 (Gottlieb). $G_1(\mathbb{R}P^n) = \begin{cases} 0 & \textit{for even } n; \\ \pi_1(\mathbb{R}P^n) & \textit{for odd } n. \end{cases}$

In view of (1.3) and $2\pi_{n-1}(SO(n)) = 0$ for n odd [10], the relation (1.52) and Corollary 2.34 yield the result [77]:

Theorem 2.36 (Pak-Woo).

$$G_n(\mathbb{R}P^n) = \gamma_{n*} G_n(\mathbb{S}^n) = \begin{cases} 0 & \textit{for even } n; \\ \pi_n(\mathbb{R}P^n) & \textit{for } n = 1, 3, 7; \\ 2\pi_n(\mathbb{R}P^n) & \textit{for odd } n \textit{ and } n \neq 1, 3, 7. \end{cases}$$

Notice that Example 2.13(1), Theorems 2.19 and 2.35, and (2.36) imply $G_k(\mathbb{R}P^n) = P_k(\mathbb{R}P^n)$ for $k = 1, n$.

Hereafter, a $\Delta_{\mathbb{F}}$-cyclic element is simply called $\mathbb{F}$-*cyclic*.

Now, we examine the equality

$$(*)_\alpha \qquad \sharp\Delta\alpha = \sharp[\iota_n, \alpha] \text{ for } \alpha \in \pi_{n+k}(\mathbb{S}^n).$$

Notice that (1.1) and Proposition 1.5 implies the divisibility $\sharp[\iota_n,\alpha] \mid \sharp\Delta\alpha$ for $\alpha \in \pi_{n+k}(\mathbb{S}^n)$ and the J-homomorphism restricted to $\{\Delta\alpha\} \subseteq \pi_{k-1}(SO(n))$ is a monomorphism if and only if the equality $(*)_\alpha$ holds.

Lemma 2.37. 1. *If* $\alpha \in \pi_{2n+k}(\mathbb{S}^{2n})$ *then* $2\Delta E\alpha = 0$.
2. *The equality* $(*)_\alpha$ *holds for* $\alpha \in \pi_{n+k}(\mathbb{S}^n)$ *provided that:*

 (i) α *is* $\mathbb{R}$*-cyclic;*
 (ii) $\sharp\alpha = \sharp[\iota_n,\alpha]$. *In particular, for* $[\iota_n,\alpha] \neq 0$ *with* $\sharp\alpha = 2$.

3. *The equality* $(*)_{E\alpha}$ *holds for* $\alpha \in \pi_{2n+k}(\mathbb{S}^{2n})$ *with* $[\iota_{2n+1}, E\alpha] \neq 0$.
4. *Let* p *be an odd prime. The equality* $(*)_{E\alpha}$ *holds for* $\alpha \in \pi_{2n+k-1}(\mathbb{S}^{2n-1}; p)$ *with* $\sharp\alpha = \sharp E^{2n}\alpha$ *for* p *an odd prime.*

Proof. 1. Let $j_n : \mathbb{R}P^{n-1} \to SO(n)$ be the reflection map. Because $q_n = p_{n+1} \circ j_{n+1}$, the relation (2.6) yields $2\iota_{2n+1} = q_{2n+1} \circ \gamma_{2n+1} = p_{2n+2} \circ j_{2n+2} \circ \gamma_{2n+1}$. This leads to $\Delta(2\iota_{2n+1}) \in \Delta p_{2n+2*}\pi_{2n+1}(SO(2n+2)) = 0$.

Now, given $\alpha \in \pi_{2n+k}(\mathbb{S}^{2n})$, the relations (2.9) yield that $2\Delta E\alpha = \Delta(2E\alpha) = \Delta((2\iota_{2n+1})E\alpha) = \Delta(2\iota_{2n+1})\alpha = 0$.
2. (i);(ii) follows directly from (1.1) and Proposition 1.5.
3. In view of (2.2) and (1.2), we obtain $2[\iota_{2n+1}, E\alpha] = 0$ for $\alpha \in \pi_{2n+k}(\mathbb{S}^{2n})$ and (1) leads to $\sharp\Delta E\alpha = \sharp[\iota_{2n+1}, E\alpha] = 2$.
4. Given $\alpha \in \pi_{2n+k-1}(\mathbb{S}^{2n-1}; p)$, owing to Lemma 1.2(2), we get $[\iota_{2n}, E\alpha] = [\iota_{2n}, \iota_{2n}] \circ E^{2n}\alpha$. Serre's isomorphism (1.7) leads to $\sharp[\iota_{2n}, E\alpha] = \sharp E^{2n}\alpha$ and, by $\sharp\alpha = \sharp E^{2n}\alpha$, (1.1), and Proposition 1.5, the result follows. □

In particular, Lemma 2.37(4) leads to:

Corollary 2.38. *The equality* $(*)_{E\alpha}$ *holds for* α *equal to* $\alpha_1(2m+1) \in \pi_{2m+4}(\mathbb{S}^{2m+1}; 3)$, $\alpha_2(2m+1) \in \pi_{2m+8}(\mathbb{S}^{2m+1}; 3)$, $\alpha_{1,5}(2m+1) \in \pi_{2m+8}(\mathbb{S}^{2m+1}; 5)$ *for* $m \geq 1$, $\alpha_{1,7}(2m+1) \in \pi_{2m+12}(\mathbb{S}^{2m+1}; 7)$ *for* $m \geq 1$, $\alpha'_3(2m+1) \in \pi_{2m+12}(\mathbb{S}^{2m+1}; 3)$ *for* $m \geq 2$, *or* $\beta_1(2m+1) \in \pi_{2m+11}(\mathbb{S}^{2m+1}; 3)$ *for* $m \geq 3$.

To state the next result, we show:

Proposition 2.39. 1. $\sharp\Delta\iota_n = \begin{cases} 1 & \textit{for } n = 1, 3, 7, \\ 2 & \textit{for odd } n \textit{ and } n \neq 1, 3, 7, \\ \infty & \textit{for even } n; \end{cases}$

2. $\sharp\Delta E\nu' = 4$;
3. $\sharp\Delta\nu_n = \begin{cases} 1 \textit{ for } n \equiv 7 \pmod 8 \textit{ or } n = 5, \\ 2 \textit{ for } n \equiv 1, 3, 5 \pmod 8 \textit{ with } n \geq 9, \\ 4 \textit{ for } n \equiv 2 \pmod 4 \textit{ with } n \geq 6 \textit{ or } n = 4, 12, \\ 8 \textit{ for } n \equiv 0 \pmod 4 \textit{ with } n \geq 8 \textit{ unless } n = 12; \end{cases}$
4. $\sharp\Delta(\nu_4\eta_7) = 2$;
5. $\sharp\Delta(E\nu')\eta_7 = 2$;
6. $\sharp\Delta(\nu_5\eta_8) = 1$;
7. $\sharp\Delta(\nu_4\eta_7^2) = 2$;
8. $\sharp\Delta(E\nu')\eta_7^2 = 2$;

9. $\sharp\Delta(\nu_5\eta_8^2) = 1$;
10. $\sharp\Delta[\iota_6, \iota_6] = 30$;
11. $\sharp\Delta\nu_n^2 = \begin{cases} 1 & \textit{for } n \equiv 4, 5, 7 \pmod 8, \\ 2 & \textit{for otherwise}; \end{cases}$
12. $\sharp\Delta\sigma''' = 1$;
13. $\sharp\Delta\sigma'' = 4$;
14. $\sharp\Delta E\sigma' = 8$;
15. $\sharp\Delta\sigma_n = \begin{cases} 16 & \textit{for } n \equiv 0 \pmod 2 \textit{ with } n \geq 10, \\ 8 & \textit{for } n = 8, \\ 1 & \textit{for } n \equiv 15 \pmod{16}, \\ 2 & \textit{for } n \equiv 1 \pmod 2 \textit{ with } n \geq 9,\ n \not\equiv 15 \pmod{16}. \end{cases}$

Proof. 1. By Lemma 2.37(2)-(ii), we have

$$\sharp\Delta\iota_n = \begin{cases} 2 & \text{for odd } n \text{ and } n \neq 1, 3, 7, \\ \infty & \text{for even } n. \end{cases}$$

2. First, in the exact sequence $(\mathcal{R}_3^4)$:

$$\pi_4(\mathbb{S}^4) \xrightarrow{\Delta} \pi_3(SO(4)) \xrightarrow{i_*} \pi_3(SO(5)) \to 0,$$

by [82, Theorem 23.6],

$$\pm\ \Delta\iota_4 = 2[\iota_3] - [\eta_2]_4. \tag{2.64}$$

Next, recall [51] that there is an isomorphism

$$\pi_k(\mathbb{S}^3) \oplus \pi_{k-1}(SO(3)) \xrightarrow{\cong} \pi_k(SO(4)) \tag{2.65}$$

given by the correspondence $(\alpha, \beta) \mapsto [\iota_3]\alpha + [\eta_2]_4\beta$.

Note that (1.20) is regarded as the J-image of (2.64).

This, in view of (2.9), implies $\pm\Delta E\nu' = \pm\Delta\iota_4 \circ \nu' = 2[\iota_3]\nu' - [\eta_2]_4\nu'$ and hence, $\sharp\Delta E\nu' = 4$.

3. We recall that

$$\pi_7(SO(5)) \cong \mathbb{Z}, \tag{2.66}$$

$\pi_k(SO(5)) = 0$ for $k = 8, 9$ and $J(\Delta\nu_4) = \pm[\iota_4, \nu_4] = \pm 2\nu_4^2$ by (1.27)(4) and Proposition 1.5. So, from the sequence $(\mathcal{R}_6^4)$, we get that

$$\Delta\nu_4 \equiv \pm[\iota_3]\nu' \pmod{[\eta_2]_4\nu'}$$

and the equality $\sharp\Delta\nu_4 = 4$ holds.

By (2.66), we have

$$\nu_5 \text{ is } \mathbb{R}\text{-cyclic.} \tag{2.67}$$

Moreover, [63, Theorem (4.3.2)] yields that

$$\nu_{8n-1} \text{ is } \mathbb{R}\text{-cyclic for } n \geq 1. \tag{2.68}$$

In the sequence $(\mathcal{R}_{8n-1}^{8n-3})$ for $n \geq 2$:

$$\pi_{8n}^{8n-3} \xrightarrow{\Delta} R_{8n-1}^{8n-3} \xrightarrow{i_*} R_{8n-1}^{8n-2},$$

we know $R_{8n-1}^{8n-3} \cong \mathbb{Z} \oplus \mathbb{Z}_2$ and $R_{8n-1}^{8n-2} \cong \mathbb{Z}$ [37, p. 161, Table]. So, we obtain

$$\sharp\Delta\nu_{8n-3} = 2.$$

By (1.31) and Lemma 2.37(1), we also get that $\sharp\Delta\nu_n = 2$ for $n \equiv 1, 3 \pmod 8$ with $n \geq 9$. In view of $(\mathcal{R}_{2n+2}^{2n})$, [37, Table, p. 161], and the fact that $\pi_{14}(SO(12)) \cong \mathbb{Z}_4 \oplus \mathbb{Z}_{24}$ [46, App. A, Table VII, Topology, p. 1745], it holds

$$\sharp\Delta\nu_{2n} = \begin{cases} 8 \text{ for } n \text{ even and } n \geq 4; \\ 4 \text{ for } n \text{ odd and } n \geq 3 \text{ or } n = 6. \end{cases}$$

4–9. By the fact that $\pi_k(SO(5)) = 0$ for $k = 8, 9$ [37, Table, p. 162] and (2.65), the map $\Delta : \pi_k(\mathbb{S}^4) \to \pi_{k-1}(SO(4)) \cong \pi_{k-1}(SO(3)) \oplus \pi_{k-1}(\mathbb{S}^3) \cong (\mathbb{Z}_2)^2$ is an isomorphism and $\Delta : \pi_{k+1}(\mathbb{S}^5) \to \pi_k(SO(5))$ is trivial. This implies $\sharp\Delta(\nu_4\eta_7) = \sharp\Delta E(\nu'\eta_6) = \sharp\Delta(\nu_4\eta_7^2) = \sharp\Delta E(\nu'\eta_6^2) = 2$ and $\sharp\Delta(\nu_5\eta_7) = \sharp\Delta(\nu_5\eta_8^2) = 1$.

10. By $(\mathcal{R}_{10}^6)$ and the fact that $\pi_{10}(SO(6)) \cong \mathbb{Z}_{120} \oplus \mathbb{Z}_2$, $\pi_{10}(SO(7)) \cong \mathbb{Z}_8$ [37, Table, p. 162], and $\pi_{10}(\mathbb{S}^6) = 0$, we obtain $\sharp\Delta[\iota_6, \iota_6] = 30$.

11. In view of (2.68), we have that ν_{8n-1}^2 is $\mathbb{R}$-cyclic for $n \geq 1$ and, by [20, Lemma 3.1], ν_{8n-k}^2 are $\mathbb{R}$-cyclic for $n \geq 1$ and $k = 3, 4$. Notice that Theorem 1.14 yields

$$\sharp\Delta\nu_n^2 = 2 \text{ for } n \not\equiv 4, 5, 7 \pmod 8 \text{ and } n \neq 2^i - 5 \text{ with } i \geq 4.$$

For $n = 2^i - 5$ with $i \geq 4$, we consider the commutative diagram

$$\begin{array}{ccccccc} & & & & \pi_{n+6}(\mathbb{S}^n) & & \\ & & & & \downarrow{\scriptstyle i'_*} & \searrow{\scriptstyle \Delta} & \\ \pi_{n+6}(SO(n)) & \xrightarrow{i_{n,n+3*}} & \pi_{n+6}(SO(n+3)) & \xrightarrow{p_*} & \pi_{n+6}(V_{n+3,3}) & \xrightarrow{\Delta'} & \pi_{n+5}(SO(n)) \\ & & & & \downarrow{\scriptstyle p'_*} & & \\ & & & & \pi_{n+6}(V_{n+3,2}), & & \end{array} \tag{2.69}$$

where the horizontal and vertical sequences are exact.

By [37, Table, p. 161], $\pi_{n+6}(SO(n+3)) \cong \mathbb{Z}_2$ for $n \geq 11$. So, in view of [14], we get $i_{n+3,\infty *} : \pi_{n+6}(SO(n+3)) \to \pi_{n+6}(SO)$ is an isomorphism. Hence, from the fact that the generator of $\pi_{n+6}(SO)$ has the $SO(6)$-of-origin [16, Theorem 1.1], we see that $i_{n,n+3*}$ is an epimorphism.

Denote by $\mathbb{R}P^n_k = \mathbb{R}P^n/\mathbb{R}P^{k-1}$ for $k \leq n$ the stunted real projective space. We obtain $\pi_{n+6}(V_{n+3,3}) \cong \pi_{n+6}(\mathbb{R}P^{n+2}_n) \cong \pi_9^s(\mathbb{R}P^5_3) \cong \pi_9^s(\mathbb{R}P^4_3) \cong \mathbb{Z}_2$ and $\pi_{n+6}(V_{n+3,2}) \cong \pi_{n+6}(\mathbb{S}^{n+1} \vee \mathbb{S}^{n+2}) = 0$. Hence, by the diagram (2.69) $\Delta\nu_n^2 \neq 0$ and so $\sharp\Delta\nu_n^2 = 2$.

12. By the relation (2.67) the element ν_5 is $\mathbb{R}$-cyclic and $\pi_8(SO(5)) = 0$ [37, Table, p. 162]. Then, (1.17)(3) and Proposition 1.12 lead to

$$\Delta(\sigma''') \in \Delta\{\nu_5, 8\iota_8, \nu_8\}_1 \subseteq \{0, 8\iota_7, \nu_7\} = \pi_8(SO(5)) \circ \nu_8 = 0.$$

13. The equation $\sharp\Delta\sigma'' = 4$ follows from (1.39).
14. First, we need the formula

$$\pm\,\Delta\iota_8 = 2[\iota_7] - [\eta_6]_8. \tag{2.70}$$

Here, we use the fact that $\pi_7(SO(7)) = \{[\eta_6]\} \cong \mathbb{Z}$ and $\pi_7(SO(8)) \cong \pi_7(\mathbb{S}^7) \oplus \pi_7(SO(7)) \cong \mathbb{Z}^2$ is generated by $[\iota_7]$ and $[\eta_6]_8$ [35, Table 2]. Next, using [35, Table 2], [36, Lemma 1.2(i)], and (2.70), we obtain $\sharp\Delta E\sigma' = 8$.
15. The relation (1.50) and Lemma 2.37(2)(ii) lead to

$$\sharp\Delta\sigma_n = 16 \text{ for } n \equiv 0 \pmod 2 \text{ with } n \geq 10.$$

Next, consider the exact sequence

$$(\mathcal{SR}^8_{14}) \quad \cdots \longrightarrow \pi_{15}(\mathbb{S}^8) \xrightarrow{\Delta} \pi_{14}(SO(8)) \xrightarrow{i_*} \pi_{14}(SO(9)) \xrightarrow{p_*} \cdots .$$

Then, applying [35, Table 2] and [36, Lemma 1.2(ii)], we get $\sharp\Delta\sigma_8 = 8$.

By [36, Theorem 1(iii) and Lemma 1.1], it holds $\sharp\Delta\sigma_{11} = 2$. In view of (1.50) and Lemma 2.37(3), we also get that $\sharp\Delta\sigma_n = 2$ for $n \equiv 1 \pmod 2$ with $n \geq 9$ and $n \neq 11$. Moreover, [63, Theorem (4.3.2)] yields that

$$\sigma_{16n-1} \text{ is } \mathbb{R}\text{-cyclic for } n \geq 1$$

and the proof is complete. □

Now, we can derive:

Corollary 2.40. *If $k \leq 7$ and $\alpha \in \pi^n_{n+k}$ then:*

1. *the equality $(*)_\alpha$ holds except $\alpha = E\nu'$, ν_{2^i-3} with $i \geq 4$, $(E\nu')\eta_7$, $\nu_4\eta_7$, $(E\nu')\eta_7^2$, $\nu_4\eta_7^2$, $[\iota_6, \iota_6]$, $\nu^2_{2^i-5}$ with $i \geq 4$, $E\sigma'$, σ_8, and σ_{11};*
2. *$\sharp\Delta\alpha = 2\sharp[\iota_n, \alpha]$ for $\alpha = E\nu'$, ν_{2^i-3} with $i \geq 4$, $(E\nu')\eta_7$, $\nu_4\eta_7$, $(E\nu')\eta_7^2$, $\nu_4\eta_7^2$, $\nu^2_{2^i-5}$ with $i \geq 4$, $E\sigma'$, and σ_{11};*
3. *$\sharp\Delta[\iota_6, \iota_6] = 10\sharp[\iota_6, [\iota_6, \iota_6]]$;*
4. *$\sharp\Delta\sigma_8^+ = 21\sharp[\iota_8, \sigma_8^+]$.*

Proof. 1. By (1.53), (1.55), (1.2)–(1.16), and Proposition 2.39(1), the equality $(*)_\alpha$ holds for $\alpha = \iota_n, \eta_n, \eta_n^2$.

The relation (1.31) and Proposition 2.39(3) yield that the equality $(*)_{\nu_n}$ holds for $n \neq 2^i - 3$ with $i \geq 4$.

Proposition 2.39(5),(8) leads to the equality $(*)_\alpha$ for $\alpha = \nu_5\eta_8, \nu_5\eta_8^2$.

In view of Theorem 1.14 and Proposition 2.39(10), the equality $(*)_{\nu_n^2}$ holds for $n \neq 2^i - 5$ with $i \geq 4$.

Lemma 2.37(2)(i)–(ii) and Proposition 2.39(11)–(12) lead to the equalities $(*)_{\sigma'''}$ and $(*)_{\sigma''}$, respectively.

By means of (1.50) and Proposition 2.39(15), the equality $(*)_{\sigma_n}$ holds for $n \neq 8, 11$.

2. The relation (1.29) and Proposition 2.39(2) yield $\sharp\Delta E\nu' = 2\sharp[\iota_4, E\nu'] = 4$.

In view of (1.31) and Proposition 2.39(3), we get $\sharp\Delta\nu_{2^i-3} = 2\sharp[\iota_{2^i-3}, \nu_{2^i-3}] = 2$ with $i \geq 4$.

By (2.2), (1.25), and (1.23), we obtain $[\iota_4, \nu_4\eta_7] = [\iota_4, (E\nu')\eta_7] = 0$ and $[\iota_4, (E\nu')\eta_7^2] = [\iota_4, \nu_4\eta_7^2] = 0$. Then, Proposition 2.39(4)–(5),(7)–(8) leads to:

$$\begin{aligned}
&\sharp\Delta(\nu_4\eta_7) = 2\sharp[\iota_4, \nu_4\eta_7] = 2,\\
&\sharp\Delta(E\nu')\eta_7 = 2\sharp[\iota_4, (E\nu')\eta_7] = 2,\\
&\sharp\Delta(\nu_4\eta_7^2) = 2\sharp[\iota_4, \nu_4\eta_7^2] = 2,\\
&\sharp\Delta(E\nu')\eta_7^2 = 2\sharp[\iota_4, (E\nu')\eta_7^2] = 2.
\end{aligned}$$

In view of Theorem 1.14 and Proposition 2.39(11), it holds $\sharp\Delta\nu_{2^i-5} = 2\sharp[\iota_{2^i-5}, \nu_{2^i-5}] = 2$ with $i \geq 4$.

The relation (1.42) and Proposition 2.39(14) yield $\sharp\Delta E\sigma' = 2\sharp[\iota_8, E\sigma'] = 8$.

By means of (1.50) and Proposition 2.39(15), it holds $\sharp\Delta\sigma_{11} = 2\sharp[\iota_{11}, \sigma_{11}] = 2$.

3. In view of the relation (1.4) and Proposition 2.39(10), we obtain $\sharp\Delta[\iota_6, \iota_6] = 10\sharp[\iota_6, [\iota_6, \iota_6]] = 30$.

4. Consider the exact sequence

$$(\mathcal{SR}_{14}^8)\quad \cdots \longrightarrow \pi_{15}(\mathbb{S}^8) \xrightarrow{\Delta} \pi_{14}(SO(8)) \xrightarrow{i_*} \pi_{14}(SO(9)) \xrightarrow{p_*} \cdots .$$

In virtue of [51, p. 132, Table], $\pi_{14}(SO(7)) \cong \mathbb{Z}_{2520} \oplus \mathbb{Z}_8 \oplus \mathbb{Z}_2$, $\pi_{14}(SO(8)) \cong \pi_{14}(SO(7)) \oplus \pi_{14}(\mathbb{S}^7) \cong \mathbb{Z}_{2520} \oplus \mathbb{Z}_{120} \oplus \mathbb{Z}_8 \oplus \mathbb{Z}_2$, and $\pi_{14}(SO(9)) \cong \mathbb{Z}_8 \oplus \mathbb{Z}_2$. Because $\Delta\nu_8^2 \neq 0$ Proposition 2.39(13), we conclude that $i_* : \pi_{14}(SO(8)) \to \pi_{14}(SO(9))$ is an epimorphism. Further, by means of Corollary 2.38 and Proposition 2.39(14), the element $\Delta E\sigma'^{+}$ generates $\mathbb{Z}_{120}$ in $\pi_{14}(SO(8))$.

Consequently, Proposition 2.39(15) yields that $\sharp\Delta\sigma_8^+ = 2520$. Then, the relation (1.51) leads to $\sharp\Delta\sigma_8^+ = 21\sharp[\iota_8, \sigma_8^+]$ and the proof is complete. □

In virtue of Proposition 2.31, $\gamma_{n*}G_k(\mathbb{S}^n)$ is an upper bound of $G_k(\mathbb{R}P^n)$ for $k \geq 2$. Now, we show the following result:

Theorem 2.41. *If* $k \leq 7$ *then the equality* $G_{k+n}(\mathbb{R}P^n) = \gamma_{n*}G_{k+n}(\mathbb{S}^n)$ *holds except the following pairs:* $(k,n) = (3,4), (3, 2^i - 3)$ *with* $i \geq 4, (4,4), (5,4), (6,4), (5,6), (6, 2^i - 5)$ *with* $i \geq 5$, $(7,8)$ *and* $(7,11)$.

Furthermore:

1. $G_7(\mathbb{R}P^4) \supseteq 12\pi_7(\mathbb{R}P^4)$;
2. $G_{2^i}(\mathbb{R}P^{2^i-3}) \supseteq 2\pi_{2^i}(\mathbb{R}P^{2^i-3})$ *for* $i \geq 4$;
3. $G_{10}(\mathbb{R}P^4) \supseteq 3\pi_{10}(\mathbb{R}P^4)$;
4. $G_{11}(\mathbb{R}P^6) \supseteq 30\pi_{11}(\mathbb{R}P^6)$;
5. $G_{15}(\mathbb{R}P^8) \supseteq 2520\pi_{15}(\mathbb{R}P^8)$;
6. $G_{18}(\mathbb{R}P^{11}) \supseteq 2\pi_{18}(\mathbb{R}P^{11})$.

Proof. Corollary 2.40 concludes that the restriction map $J \mid_{\Delta\pi_{n+k}(\mathbb{S}^n)}: \Delta\pi_{n+k}(\mathbb{S}^n) \to \pi_{k+2n-1}(\mathbb{S}^n)$ is not a monomorphism if and only if $(k, n) = (3, 4), (3, 2^i - 3)$ with $i \geq 4, (4, 4), (5, 4), (6, 4), (5, 6), (6, 2^i - 5)$ with $i \geq 5, (7, 8)$ and $(7, 11)$. In the light of Lemmas 2.2 and 2.33, Proposition 2.16(1), Chap. 1, and [82], we obtain the equality.

By Proposition 2.39(3), we get $\sharp\Delta\nu_4^+ = 12$. This and (2.61) imply (1): $G_7(\mathbb{R}P^4) \supseteq p'_*\pi_7(SO(5)) = 12\pi_7(\mathbb{R}P^4)$.

(2) follows from the fact that Ker $\{\Delta : \pi_{2^i}(\mathbb{S}^{2^i-3}) \to \pi_{2^i-1}(SO(2^i - 3))\} = 2\pi_{2^i}(\mathbb{S}^{2^i-3})$ (Proposition 2.39(3)).

In view of Proposition 2.39(11), it holds $\sharp\Delta(\nu_4^+)^2 = 3$. This and (2.61) imply (3): $G_{10}(\mathbb{R}P^4) \supseteq p'_*\pi_{10}(SO(5)) = 3\pi_{10}(\mathbb{R}P^4)$. By means of Proposition 2.39(10), we know that $\sharp\Delta[\iota_6, \iota_6] = 30$. Hence, Lemma 2.33(1) yields (4).

By Corollary 2.40(4), it holds

$$\mathrm{Ker}\{\Delta : \pi_{15}(\mathbb{S}^8) \to \pi_{14}(SO(8))\} = 2520\pi_{15}(\mathbb{S}^8)$$

and this leads to (5).

(6) follows from the fact that Ker $\{\Delta : \pi_{18}(\mathbb{S}^{11}) \to \pi_{17}(SO(11))\} = 2\pi_{18}(\mathbb{S}^{11})$ (Proposition 2.39(15)). This completes the proof. □

To state the next result, we recall that

$$\varepsilon_{4n+3} \text{ and } \mu_{4n+3} \text{ are } \mathbb{R}\text{-cyclic [20, Example 3.2].} \tag{2.71}$$

By Lemma 2.37(1), we get

$$\Delta E\pi_{2n+k}(\mathbb{S}^{2n}; p) = 0 \text{ for an odd prime } p. \tag{2.72}$$

Proposition 2.42. $G_{n+k}(\mathbb{R}P^n) = \pi_{n+k}(\mathbb{R}P^n)$ *if* $n \equiv 3 \pmod 4$ *and* $k = 8, 9$ *or* $n \equiv 3 \pmod 4$ *with* $n \geq 7$ *and* $k = 10$, *or* $n \equiv 1 \pmod 4$, $n \equiv 7 \pmod 8$ *with* $n \geq 13$ *and* $k = 11$, *or* $n \equiv 1 \pmod 2$ *with* $n \geq 15$ *and* $k = 13$.

Proof. By (2.9) and (1.53), we get $\Delta(\eta_n\sigma_{n+1}) = \Delta(\eta_n\mu_{n+1}) = 0$ for $n \equiv 3 \pmod 4$. This and (2.71) imply $\Delta\pi_{n+k}(\mathbb{S}^n) = \Delta\pi^n_{n+10} = 0$ for $k = 8, 9$ and $n \equiv 3 \pmod 4$. By (2.72), $\Delta\pi_{n+10}(\mathbb{S}^n; 3) = 0$ for n odd with $n \geq 7$.

Next, by [30, Table 2] and the diagram in [20, p. 402], $\Delta\zeta_n = 0$ for $n \equiv 1 \pmod 4$, $n \equiv 7 \pmod 8$, and $n \geq 13$. This, Corollary 2.34, and (2.72) imply

$\gamma_{n*}\pi_{n+11}(\mathbb{S}^n) \subseteq G_{n+11}(\mathbb{R}P^n)$ for $n \equiv 1 \pmod 4$, $n \equiv 7 \pmod 8$ with $n \geq 13$. Because $\pi_{n+13}(\mathbb{S}^n) \cong \mathbb{Z}_3$ for $n \geq 15$, in view of (2.72), we have $\Delta\pi_{n+13}(\mathbb{S}^n) = 0$ for $n \equiv 1 \pmod 2$ with $n \geq 15$. Then, by means of Corollary 2.34, the proof is complete. □

2.7 Gottlieb Groups of Complex and Quaternionic Projective Spaces

By [23, Theorem 1], we obtain

$$G_2(\mathbb{C}P^n) = 0 \text{ for } n \geq 1$$

proved in [19, 39] and [40] as well.

To consider $G_k(\mathbb{C}P^n)$ for $k > 2$, we recall the group structures of $\pi_{2n+k-1}(SU(n))$ for $0 \leq k \leq 7$ from [13, 14, 37], [46, 47, App. A, Table VII, Topology, pp. 1745–1746], and [48].

$$\pi_{2n-1}(SU(n)) \cong \mathbb{Z};\ \pi_{2n}(SU(n)) \cong \mathbb{Z}_{n!} \text{ for } n \geq 2;$$

$$\pi_{2n+1}(SU(n)) \cong \begin{cases} \mathbb{Z}_2 & \text{for even } n, \\ 0 & \text{for odd } n; \end{cases}$$

$$\pi_{2n+2}(SU(n)) \cong \begin{cases} \mathbb{Z}_{(n+1)!} \oplus \mathbb{Z}_2 & \text{for even } n \geq 4, \\ \mathbb{Z}_{\frac{(n+1)!}{2}} & \text{for odd } n \geq 3; \end{cases}$$

$$\pi_{2n+3}(SU(n)) \cong \begin{cases} \mathbb{Z}_{(24,n)} & \text{for even } n, \\ \mathbb{Z}_{\frac{(24,n+3)}{2}} & \text{for odd } n \geq 3; \end{cases}$$

$$\pi_{2n+4}(SU(n)) \cong \begin{cases} \mathbb{Z}_{\frac{(n+2)!(24,n)}{48}} & \text{for even } n \geq 4, \\ \mathbb{Z}_{\frac{(n+2)!(24,n+3)}{24}} & \text{for odd } n \geq 3; \end{cases}$$

$$\pi_{2n+5}(SU(n)) \cong \pi_{2n+5}(SU(n+1)) \cong \begin{cases} \mathbb{Z}_{\frac{(24,n+4)}{2}} & \text{for even } n, \\ \mathbb{Z}_{(24,n+1)} & \text{for odd } n \geq 3; \end{cases}$$

$$\pi_{2n+6}(SU(n)) \cong \begin{cases} \pi_{2n+6}(SU(n+1)) & \text{for } n \equiv 2, 3 \pmod 4 \text{ with } n \geq 3, \\ \pi_{2n+6}(SU(n+1)) \oplus \mathbb{Z}_2 & \text{for } n \equiv 0, 1 \pmod 4 \text{ with } n \geq 4. \end{cases}$$

Write $\omega_n = \omega_{n,\mathbb{C}}$ and notice that $\Delta_{\mathbb{C}}\iota_{2n+1} = \omega_n = xj_n E\gamma_{n-1}$ with $(x, n!) = 1$ is a generator of $\pi_{2n}(SU(n))$, where $j_n : E\mathbb{C}P^{n-1} \hookrightarrow SU(n)$ is the canonical inclusion satisfying $p_n j_n = Eq_{n-1}$. Further, by (2.6), we get

$$p_n\omega_n = (n-1)\eta_{2n-1}. \tag{2.73}$$

In view of [60, Proposition 6.5 (i)], there exists such $\varrho_n \in \pi_{2n+2}(E\mathbb{C}P^{n-1})$ that

$$(Ei''_n)\varrho_n = \begin{cases} E\gamma_n & \text{for even } n, \\ 2E\gamma_n & \text{for odd } n, \end{cases}$$

where $i''_n : \mathbb{C}P^{n-1} \hookrightarrow \mathbb{C}P^n$ is the inclusion map.

To search a generator of $\pi_{2n+4}(SU(n))$, notice that, by means of (2.73), the result [60, Proposition 7.3 i)] yields the trivial map $i_{n+2*} : \pi_{2n+5}(SU(n+1)) \longrightarrow \pi_{2n+5}(SU(n+2))$. Hence, $i_{n,n+2,*} : \pi_{2n+5}(SU(n)) \longrightarrow \pi_{2n+5}(SU(n+2))$ is trivial as well. Then, the fibration $p''_{n+2} : SU(n+2) \xrightarrow{SU(n)} SU(n+2)/SU(n)$ leads to a monomorphism $p''_{n+2*} : \pi_{2n+5}(SU(n+2)) \longrightarrow \pi_{2n+5}(SU(n+2)/SU(n))$. Because, in view of [18], it holds

$$\pi_{2n+5}(SU(n+2)/SU(n)) \cong \begin{cases} \mathbb{Z}_2 & \text{for even } n \geq 4, \\ 0 & \text{for odd } n \geq 3, \end{cases}$$

we deduce that $p''_{n+2*} : \pi_{2n+5}(SU(n+2)) \longrightarrow \pi_{2n+5}(SU(n+2)/SU(n))$ is an isomorphism for $n \geq 3$. Hence, $i_{n,n+2*} : \pi_{2n+4}(SU(n)) \longrightarrow \pi_{2n+4}(SU(n+2))$ is a monomorphism. Consequently, there exists such a generator $\omega'_n \in \pi_{2n+4}(SU(n))$ that

$$i_{n,n+2}\omega'_n = \begin{cases} \frac{48}{(24,n)}\omega_{n+2} & \text{for even } n \geq 4, \\ \frac{24}{(24,n+3)}\omega_{n+2} & \text{for odd } n \geq 3. \end{cases}$$

The exact sequences $(\mathcal{SC}^n_{2n+k})$ for $k = 5, 6$ lead to the isomorphism

$$i_* : \pi_{2n+5}(SU(n)) \xrightarrow{\cong} \pi_{2n+5}(SU(n+1)$$

and the epimorphism

$$\pi_{2n+6}(SU(n)) \xrightarrow{i_*} \pi_{2n+6}(SU(n+1)) \to 0,$$

respectively. Hence, there exist such generators $\omega''_n \in \pi_{2n+5}(SU(n))$ and $\omega'''_n \in \pi_{2n+6}(SU(n))$ that $i_{n+1}\omega''_n = \omega_{n+1}\nu^+_{2n+2}$ and $i_{n+1}\omega'''_n = \omega'_{n+1}$ for $n > 2$.

Making use of the results above and [60, Propositions 3.2, 7.2, and 7.3], we obtain the following result which overlaps with those of [37, 47, pp. 163–165] and [48]:

Lemma 2.43.

$$\pi_{2n+k}(SU(n)) = \begin{cases} \{\omega_n\} \cong \mathbb{Z}_{n!} & \textit{for } k = 0 \textit{ and } n \geq 2, \\ \{\omega_n \eta_{2n}\} \cong \mathbb{Z}_2 & \textit{for } k = 1 \textit{ and even } n, \\ \{j_n \varrho_n, \omega_n \eta_{2n}^2\} \cong \mathbb{Z}_{(n+1)!} \oplus \mathbb{Z}_2 & \textit{for } k = 2 \textit{ and even } n \geq 4, \\ \{j_n \varrho_n\} \cong \mathbb{Z}_{\frac{(n+1)!}{2}} & \textit{for } k = 2 \textit{ and odd } n \geq 3, \\ \{\omega_n \nu_{2n}^+\} \cong \mathbb{Z}_{(24,n)} & \textit{for } k = 3 \textit{ and even } n, \\ \{\omega_n \nu_{2n}^+\} \cong \mathbb{Z}_{\frac{(24,n+3)}{2}} & \textit{for } k = 3 \textit{ and odd } n \geq 3, \\ \{\omega_n'\} \cong \mathbb{Z}_{\frac{(n+2)!(24,n)}{48}} & \textit{for } k = 4 \textit{ and even } n, \\ \{\omega_n'\} \cong \mathbb{Z}_{\frac{(n+2)!(24,n+3)}{24}} & \textit{for } k = 4 \textit{ and odd } n \geq 3, \\ \{\omega_n''\} \cong \mathbb{Z}_{\frac{(24,n+4)}{2}} & \textit{for } k = 5 \textit{ and even } n \geq 4, \\ \{\omega_n''\} \cong \mathbb{Z}_{(24,n+1)} & \textit{for } k = 5 \textit{ and odd } n \geq 3, \end{cases}$$

$$\begin{cases} \{\omega_n''', \omega_n \nu_{2n}^2\} \cong \mathbb{Z}_{\frac{(n+3)!(24,n+4)}{24}} \oplus \mathbb{Z}_2 & \textit{for } k = 6 \textit{ and } n \equiv 0 \pmod 4 \textit{ with } n \geq 4, \\ \{\omega_n''', \omega_n \nu_{2n}^2\} \cong \mathbb{Z}_{\frac{(n+3)!(24,n+1)}{48}} \oplus \mathbb{Z}_2 & \textit{for } k = 6 \textit{ and } n \equiv 1 \pmod 4 \textit{ with } n \geq 5, \\ \{\omega_n'''\} \cong \mathbb{Z}_{\frac{(n+3)!(24,n+4)}{24}} & \textit{for } k = 6 \textit{ and } n \equiv 2 \pmod 4 \textit{ with } n \geq 4, \\ \{\omega_n'''\} \cong \mathbb{Z}_{\frac{(n+3)!(24,n+1)}{48}} & \textit{for } k = 6 \textit{ and } n \equiv 3 \pmod 4 \textit{ with } n \geq 3. \end{cases}$$

Further:

1. $\sharp \Delta_{\mathbb{C}} \iota_{2n+1} = n!$;
2. $\sharp \Delta_{\mathbb{C}} \eta_{2n+1} = \begin{cases} 2 \textit{ for even } n, \\ 1 \textit{ for odd } n; \end{cases}$
3. $\sharp \Delta_{\mathbb{C}} \eta_{2n+1}^2 = \begin{cases} 2 \textit{ for even } n, \\ 1 \textit{ for odd } n; \end{cases}$
4. $\sharp \Delta_{\mathbb{C}} \nu_{2n+1}^+ = \begin{cases} (24, n) & \textit{for even } n, \\ (24, n+3)/2 & \textit{for odd } n \geq 3; \end{cases}$
5. $\sharp \Delta_{\mathbb{C}} \nu_{2n+1}^2 = \begin{cases} 1 \textit{ for } n \equiv 2, 3 \pmod 4 \textit{ and } n \geq 2, \\ 2 \textit{ for } n \equiv 0, 1 \pmod 4 \textit{ and } n \geq 4. \end{cases}$

By Lemmas 2.33(1) and 2.43(1), we obtain [39, Theorem III.8]:

Theorem 2.44 (Lang). $n!\pi_{2n+1}(\mathbb{C}P^n) \subseteq G_{2n+1}(\mathbb{C}P^n)$.

By Theorems 2.20(1) and 2.44, it holds

$$G_5(\mathbb{C}P^2) = P_5(\mathbb{C}P^2) = 2\pi_5(\mathbb{C}P^2).$$

In view of Lemma 2.43(2),(4), it holds:

$$\begin{aligned} &(1)\ \eta_{4n+3} \text{ and} \\ &(2)\ (12, n+2)\nu_{4n+3}^+ \end{aligned} \tag{2.74}$$

are $\mathbb{C}$-cyclic.

Now, we show:

Theorem 2.45. 1. *Let* $k = 1, 2$. *Then*

$$G_{k+2n+1}(\mathbb{C}P^n) = \begin{cases} 0 & \text{for even } n, \\ \pi_{2n+k+1}(\mathbb{C}P^n) \cong \mathbb{Z}_2 & \text{for odd } n. \end{cases}$$

2. $G_{2n+4}(\mathbb{C}P^n) \supseteq \begin{cases} (24,n)\pi_{2n+4}(\mathbb{C}P^n) \cong \mathbb{Z}_{\frac{24}{(24,n)}} & \text{for even } n, \\ \frac{(24,n+3)}{2}\pi_{2n+4}(\mathbb{C}P^n) \cong \mathbb{Z}_{\frac{48}{(24,n+3)}} & \text{for odd } n. \end{cases}$
In particular, $G_{2n+4}(\mathbb{C}P^n) = 2\pi_{2n+4}(\mathbb{C}P^n)$ *for* $n \equiv 2, 10 \pmod{12}$ *with* $n \geq 10$ *except* $n = 2^{i-1} - 2$ *or* $n \equiv 1, 17 \pmod{24}$ *for* $n \geq 17$ *and* $G_{2n+4}(\mathbb{C}P^n) = \pi_{2n+4}(\mathbb{C}P^n)$ *for* $n \equiv 7, 11 \pmod{12}$.
3. $G_{2n+6}(\mathbb{C}P^n) = \pi_{2n+6}(\mathbb{C}P^n) \cong \begin{cases} 0 & \text{for } n \geq 3, \\ \mathbb{Z}_2 & \text{for } n = 2. \end{cases}$
4. $G_{2n+7}(\mathbb{C}P^n) = \pi_{2n+7}(\mathbb{C}P^n)$ *for* $n \equiv 2, 3 \pmod 4$.

Proof. By Proposition 2.16(1) and the fact that $G_{k+2n+1}(\mathbb{S}^{2n+1}) = 0$ if $k = 1, 2$ and n is even, we have $G_{k+2n+1}(\mathbb{C}P^n) = 0$ in this case. By Lemmas 2.33(1) and 2.43(2) and (3), $G_{2n+k+1}(\mathbb{C}P^n) = \pi_{2n+k+1}(\mathbb{C}P^n)$ for n odd. This leads to (1).

In view of Corollary 2.34, Lemmas 2.33(1) and 2.43(4) imply the first half of (2).

As it is easily seen (1.31), for n even, $(24, n) = \sharp[\iota_{2n+1}, \nu^+_{2n+1}] = 2$ if and only if $n \equiv 2, 10 \pmod{12}$ for $n \geq 10$ and $n \neq 2^{i-1} - 2$. For n odd, $\frac{(24,n+3)}{2} = \sharp[\iota_{2n+1}, \nu^+_{2n+1}] = 2$ if and only if $n \equiv 1, 17 \pmod{24}$ for $n \geq 17$. Moreover, $\frac{(24,n+3)}{2} = \sharp[\iota_{2n+1}, \nu^+_{2n+1}] = 1$ if and only if $n \equiv 7, 11 \pmod{12}$. This and Lemma 2.33(2) lead to the second half of (2).

Since $\Delta_{\mathbb{C}} : \pi_{10}(\mathbb{S}^5) \to \pi_9(SU(2)) \cong \mathbb{Z}_3$ is trivial, we have $G_{10}(\mathbb{C}P^2) = \pi_{10}(\mathbb{C}P^2) \cong \mathbb{Z}_2$ and this leads to (3).

(4) follows from Lemmas 2.33(1) and 2.43(5), and the proof is complete. □

Next, we present:

Proposition 2.46. $G_k(\mathbb{C}P^3; 2) = \pi_k(\mathbb{C}P^3; 2)$ *for* $8 \leq k \leq 24$ *unless* $k = 15, 21$. *Further,* $\gamma_{3,\mathbb{C}*}\{\sigma'\eta_{14}, \bar{\nu}_7 + \varepsilon_7\} \subseteq G_{15}(\mathbb{C}P^3; 2)$ *and* $\gamma_{3,\mathbb{C}*}\{\sigma'\sigma_{14}\} \subseteq G_{21}(\mathbb{C}P^3; 2)$.

Proof. We make use of Lemma 2.33(1) and show that $\Delta_{\mathbb{C}}|_{\pi^7_{k+1}} : \pi^7_{k+1} \to \pi_k(SU(3); 2)$ is trivial for $7 \leq k \leq 22$ unless $k = 15, 21$. First, we recall:

$$\pi_k(\mathbb{S}^7) = \begin{cases} \{\eta_7\} \cong \mathbb{Z}_2 & \text{for } k = 8; \\ \{\eta_7^2\} \cong \mathbb{Z}_2 & \text{for } k = 9; \\ \{\nu_7^+\} \cong \mathbb{Z}_{24} & \text{for } k = 10; \\ 0 & \text{for } k = 11; \\ 0 & \text{for } k = 12; \\ \{\nu_7^2\} \cong \mathbb{Z}_2 & \text{for } k = 13; \\ \{\sigma'^+\} \cong \mathbb{Z}_{120} & \text{for } k = 14; \\ \{\sigma'\eta_{14}, \bar{\nu}_7, \varepsilon_7\} \cong (\mathbb{Z}_2)^3 & \text{for } k = 15; \\ \{\sigma'\eta_{14}^2, \nu_7^3, \mu_7, \eta_7\varepsilon_8\} \cong (\mathbb{Z}_2)^4 & \text{for } k = 16; \\ \{\nu_7\sigma_{10}, \eta_7\mu_8, \beta_1(7)\} \cong \mathbb{Z}_{24} \oplus \mathbb{Z}_2 & \text{for } k = 17; \\ \{\zeta_7^+, \bar{\nu}_7\nu_{15}\} \cong \mathbb{Z}_{504} \oplus \mathbb{Z}_2 & \text{for } k = 18; \\ 0 & \text{for } k = 19; \end{cases} \tag{2.75}$$

$$\begin{cases} \{\nu_7\sigma_{10}\nu_{17}, \alpha_1(7)\beta_1(10)\} \cong \mathbb{Z}_6 & \text{for } k = 20; \\ \{\sigma'\sigma_{14}, \kappa_7, \alpha'\} \cong \mathbb{Z}_{24} \oplus \mathbb{Z}_4 & \text{for } k = 21; \\ \{\rho''^+, \sigma'\bar{\nu}_{14}, \sigma'\varepsilon_{14}, \bar{\varepsilon}_7\} \cong \mathbb{Z}_{120} \oplus (\mathbb{Z}_2)^3 & \text{for } k = 22; \\ \{\sigma'\mu_{14}, E\zeta', \mu_7\sigma_{16}, \eta_7\bar{\varepsilon}_8\} \cong (\mathbb{Z}_2)^4 & \text{for } k = 23; \\ \{\sigma'\eta_{14}\mu_{15}, \nu_7\kappa_{10}, \bar{\mu}_7, \eta_7\mu_8\sigma_{17}\} \cong (\mathbb{Z}_2)^4 & \text{for } k = 24, \end{cases}$$

where $E\alpha' = [[\iota_8, \iota_8], \iota_8]$.

Further, we make use of the relation

$$\Delta_{\mathbb{C}} = i_{3,\mathbb{C}*}\Delta_{\mathbb{H}} \tag{2.76}$$

determined by (1.48)(2). Since $\pi_6(Sp(2)) = \{\omega\}$ and $\pi_6(Sp(2)) = 0$ [55, Theorem 5.1], we have $\Delta_{\mathbb{H}}(\iota_7) = x\omega$ for an integer x with $(x, 12) = 1$. Hence, (2.76) yields

$$\Delta_{\mathbb{C}}(\iota_7) = x i_{3,\mathbb{C}*}\omega. \tag{2.77}$$

But, $\pi_4(SU(3)) = 0$ [54, Theorem 4.1] leads to $i_{3,\mathbb{C}*}\eta_3 = 0$. Then, the relations (2.9), (1.33), and (2.77) imply $\Delta_{\mathbb{C}}(\eta_7) = i_{3,\mathbb{C}*}(\omega\eta_6) = (i_{3,\mathbb{C},*}\eta_3)\nu_4 = 0$. Next, (2.9) and (1.28) lead to $\Delta_{\mathbb{C}}(\nu_7) = i_{3,\mathbb{C}*}(\omega\nu_6) = i_{3,\mathbb{C}*}(\nu'\nu_6) = 0$. Consequently, in view of (2.9), $\Delta_{\mathbb{C}}(\eta_7 E\alpha) = \Delta_{\mathbb{C}}(\nu_7 E\beta) = 0$ for any $\alpha \in \pi_n^7$ and $\beta \in \pi_n^9$.

Because $\pi_8(SU(3); 2) \cong \mathbb{Z}_4$ [54, Theorem 4.1], using the fact that $\Delta_{\mathbb{C}}(\nu_7^3) = 0$, the relations (1.35) and (1.60) and Proposition 1.12 yield $\Delta_{\mathbb{C}}(\mu_7) \in \Delta_{\mathbb{C}}\{\eta_7, 2\iota_8, E^3\sigma'''\}_3 \subseteq \{0, 2\iota_7, E^2\sigma'''\}_2 = \pi_8(SU(3); 2) \circ 4E\sigma' = 0$. Hence, (2.9) implies $\Delta_{\mathbb{C}}(\mu_7 E\delta) = 0$ for any $\delta \in \pi_n^{15}$.

In view of $\pi_{16}(SU(3); 2) \cong \mathbb{Z}_4 \oplus \mathbb{Z}_2$ [85, Lemma 5.14], (2.31), and Proposition 1.12, we get $\Delta_{\mathbb{C}}(\bar{\mu}_7) \in \Delta_{\mathbb{C}}\{\mu_7, 2\iota_{16}, 8\sigma_{16}\}_5 \subseteq \{0, 2\iota_{15}, 8\sigma_{15}\}_4 = \pi_{16}(SU(3); 2) \circ 8\sigma_{16} = 0$. Further, (2.20) and (2.77) lead to $\Delta_{\mathbb{C}}(\zeta_7) = 0$.

Because $\Delta_{\mathbb{H}}(\sigma') = 2\varepsilon'$, $\Delta_{\mathbb{H}}(\kappa_7) = \pm\bar{\varepsilon}'$ and $\Delta_{\mathbb{H}}(\rho'') = 0$ [54, Proposition 3.2 ii)], we apply [54, Theorem 4.1] and (2.76) to get $\Delta_{\mathbb{C}}(\sigma') = 2i_{3,\mathbb{C}*}(\varepsilon') = 0$,

$\Delta_{\mathbb{C}}(\kappa_7) = i_{3,\mathbb{C}*}(\bar{\varepsilon}') \neq 0$, and $\Delta_{\mathbb{C}}(\rho'') = 0$. Further, (2.9) yields $\Delta_{\mathbb{C}}(\sigma' E\gamma) = 0$ for any $\gamma \in \pi_n^{13}$. In view of (1.106), the above leads to $\Delta_{\mathbb{H}}(E\zeta') = \nu'\zeta' = 2\varepsilon'\eta_{13}\varepsilon_{14} = 0$ and $\Delta_{\mathbb{C}}(E\zeta') = 0$.

Next, (2.9), (1.122)(1), (2.77), and [54, (4.1)] imply $\Delta_{\mathbb{C}}(\varepsilon_7) = \Delta_{\mathbb{C}}(\bar{\nu}_7) = i_{3,\mathbb{C}*}(\nu'\varepsilon_6) = i_{3,\mathbb{C}*}(\varepsilon_3\nu_{11}) = 2([\nu_5^2] \circ \nu_{11}) \neq 0$ and $\Delta_{\mathbb{C}}(\bar{\nu}_7 + \varepsilon_7) = 0$. Finally, (1.123) leads to $\Delta_{\mathbb{C}}(\bar{\nu}_7\nu_{15}) = i_{3\mathbb{C}*}(\nu'\varepsilon_6\nu_{14}) = 0$ and the proof follows. □

Then, we obtain:

Corollary 2.47. 1. *$G_k(\mathbb{C}P^2) = \pi_k(\mathbb{C}P^2)$ for $10 \leq k \leq 12$;*
2. *$G_k(\mathbb{C}P^2; p) = \pi_k(\mathbb{C}P^2; p)$ for an odd prime p;*
3. *$G_k(\mathbb{C}P^{2n+1}) \supseteq \gamma_{2n+1}\eta_{4n+3}^m \circ \pi_k^{4n+m+3}$ for $m = 1, 2$;*
4. (i) *$G_k(\mathbb{C}P^{2n}) \supseteq 2(12, n)\gamma_{2n}\nu_{4n+1}^+ \circ \pi_k(\mathbb{S}^{4n+4})$;*
(ii) *$G_k(\mathbb{C}P^{2n+1}) \supseteq (12, n+2)\gamma_{2n+1}\nu_{4n+3}^+ \circ \pi_k(\mathbb{S}^{4n+6})$;*
5. *$G_k(\mathbb{C}P^{4n+2}) \supseteq \gamma_{4n+2}\nu_{8n+5}^2 \circ \pi_k^{8n+11}$;*
6. *$G_{8n+11}(\mathbb{C}P^{4n+1}) = \pi_{8n+11}(\mathbb{C}P^{4n+1})$ for $n \geq 2$;*
7. *$G_{8n+k}(\mathbb{C}P^{4n+3}) = \pi_{8n+k}(\mathbb{C}P^{4n+3})$ for $k = 28, 29$ with $n \geq 2$.*

Proof. By [54, Lemma 3.1(i)], we get $\Delta_{\mathbb{C}}\pi_k(\mathbb{S}^5) = 0$ for $10 \leq k \leq 12$. Hence, Lemma 2.33(1) yields (1).

Corollary 2.2 and Theorem 2.45 yield (2)–(5).

The relations (1.57), (2.9), and (2.74)(1) lead to

$$\begin{aligned}\Delta_{\mathbb{C}}\varepsilon_{4n+3} &= (\Delta_{\mathbb{C}}\iota_{4n+3}) \circ \varepsilon_{4n+2} \in \{\omega_{2n+1}, \eta_{4n+2}, \nu_{4n+3}\} \circ 2\nu_{4n+7} \\ &\subseteq \pi_{4n+7}(SU(2n+1)) \circ 2\nu_{4n+7}.\end{aligned}$$

Then, by the group structure $\pi_{4n+7}(SU(2n+1)) \cong \mathbb{Z}_{2(12,n+1)}$, we obtain that ε_{4n+3} is $\mathbb{C}$-cyclic for even n. Hence, by (2.9), (1.68), and (2.74), the element $\bar{\nu}_{4n+3}$ is $\mathbb{C}$-cyclic for even n as well. Because $\pi_8^s = \{\bar{\nu}, \varepsilon\} \cong (\mathbb{Z}_2)^2$ [85, Theorem 7.1], this yields (6).

Notice that (2.9) and (2.74) imply the $\mathbb{C}$-cyclicity of the elements: $\eta_{8n+7}\bar{\kappa}_{8n+8}$, $\nu_{8n+7}\xi_{8n+10}$, $\eta_{8n+7}^2\bar{\kappa}_{8n+8}$, and $\nu_{8n+7}\bar{\sigma}_{8n+10}$. Further, recall from [49] that $\pi_{21}^s = \{\eta\bar{\kappa}, \sigma^3\} \cong (\mathbb{Z}_2)^2$ and $\pi_{22}^s = \{\varepsilon\kappa, \nu\bar{\sigma}\} \cong (\mathbb{Z}_2)^2$. Then, (2.33) and (2.27) lead to (7) for $k = 28, 29$ and $n \geq 2$, and the proof is complete. □

Now, we consider the case of the quaternionic projective space. By [23, Theorem 1], we obtain

$$G_4(\mathbb{H}P^n) = 0 \quad \text{for} \quad n \geq 1$$

proved in [19, 39] and [40] as well.

By Proposition 2.30(1);(4) and Chap. 1,

$$G_k(\mathbb{H}P^n) = 0 \quad \text{for } n \geq 1 \text{ and } k = 5, 6;$$

$$G_k(\mathbb{H}P^n) = 0 \quad \text{for } n \geq 3 \text{ and } k = 12, 13.$$

We recall the group structures $\pi_{4n+k-1}(Sp(n))$ for $-1 \le k \le 6$ from [14], [46, App. A, Table VII, Topology, p. 1746], [50], and [55]:

$$\pi_{4n-2}(Sp(n)) = 0;\ \pi_{4n-1}(Sp(n)) \cong \mathbb{Z};$$

$$\pi_{4n}(Sp(n)) \cong \begin{cases} 0 & \text{for even } n, \\ \mathbb{Z}_2 & \text{for odd } n; \end{cases}$$

$$\pi_{4n+1}(Sp(n)) \cong \begin{cases} 0 & \text{for even } n, \\ \mathbb{Z}_2 & \text{for odd } n; \end{cases}$$

$$\pi_{4n+2}(Sp(n)) \cong \begin{cases} \mathbb{Z}_{(2n+1)!} & \text{for even } n, \\ \mathbb{Z}_{2(2n+1)!} & \text{for odd } n; \end{cases}$$

$$\pi_{4n+3}(Sp(n)) \cong \mathbb{Z}_2;$$

$$\pi_{4n+4}(Sp(n)) \cong \begin{cases} (\mathbb{Z}_2)^2 & \text{for even } n, \\ \mathbb{Z}_2 & \text{for odd } n; \end{cases}$$

$$\pi_{4n+5}(Sp(n)) \cong \begin{cases} \mathbb{Z}_{(24,n+2)} \oplus \mathbb{Z}_2 & \text{for even } n; \\ \mathbb{Z}_{(24,n+2)} & \text{for odd } n. \end{cases}$$

Write θ_{4n-1} for a generator on $\pi_{4n-1}(Sp(n))$. Then, [55] yields

$$p_*(\theta_{4n+3}) = \begin{cases} (2n+1)!\iota_{4n+3} & \text{for even } n, \\ 2(2n+1)!\iota_{4n+3} & \text{for odd } n. \end{cases}$$

Denoting $\omega_n = \omega_{n,\mathbb{H}}$ and making use of the exact sequences $(\mathcal{SH}^n_{4n+k})$ for $k = 2, 3, 4, 5$, (2.9), [50, 55], [85, Proposition 1.8], and the results above, we obtain:

Lemma 2.48.

$$\pi_{4n+k}(Sp(n)) = \begin{cases} \{\theta_{4n-1}\eta_{4n-1}\} \cong \mathbb{Z}_2 & \textit{for } k = 0 \textit{ and odd } n, \\ \{\theta_{4n-1}\eta^2_{4n-1}\} \cong \mathbb{Z}_2 & \textit{for } k = 1 \textit{ and odd } n, \\ \{\omega_n\} \cong \mathbb{Z}_{(2n+1)!} & \textit{for } k = 2 \textit{ and even } n, \\ \{\omega_n\} \cong \mathbb{Z}_{2(2n+1)!} & \textit{for } k = 2 \textit{ and odd } n, \\ \{\omega_n\eta_{4n+2}\} \cong \mathbb{Z}_2 & \textit{for } k = 3 \textit{ and any } n, \\ \{\omega_n\eta^2_{4n+2}, \alpha\} \cong (\mathbb{Z}_2)^2 & \textit{for } k = 4 \textit{ and even } n, \\ \{\omega_n\eta^2_{4n+2}\} \cong \mathbb{Z}_2 & \textit{for } k = 4 \textit{ and odd } n, \\ \{\omega_n\nu^+_{4n+2}, \alpha\eta_{4n+4}\} \cong \mathbb{Z}_{(24,n+2)} \oplus \mathbb{Z}_2 & \textit{for } k = 5 \textit{ and even } n, \\ \{\omega_n\nu^+_{4n+2}\} \cong \mathbb{Z}_{(24,n+2)} & \textit{for } k = 5 \textit{ and odd } n, \end{cases}$$

where $\alpha \in -\{\omega_n, (2n+1)!\iota_{4n+2}, \eta_{4n+2}\}$ *with* $i_*(\alpha) = \theta_{4n-1} \circ \eta_{4n-1}$.

Further:

1. $\sharp\Delta_{\mathbb{H}}\iota_{4n+3} = \begin{cases} (2n+1)! & \textit{for even } n, \\ 2(2n+1)! & \textit{for odd } n; \end{cases}$
2. $\sharp\Delta_{\mathbb{H}}\eta^k_{4n+3} = 2$ *for* $k = 1, 2$;
3. $\sharp\Delta_{\mathbb{H}}\nu^+_{4n+3} = (24, n+2)$.

By Lemma 2.48(3), we get that

$$(24, n+2)\nu^+_{4n+3} \text{ is } \mathbb{H}\text{-cyclic.} \tag{2.78}$$

By Lemmas 2.2, 2.33(1), and 2.48(1);(3), we obtain:

Theorem 2.49. 1. $G_{4n+3}(\mathbb{H}P^n) \supseteq \begin{cases} (2n+1)!\gamma_{n*}\pi_{4n+3}(\mathbb{S}^{4n+3}) & \textit{for even } n, \\ 2(2n+1)!\gamma_{n*}\pi_{4n+3}(\mathbb{S}^{4n+3}) & \textit{for odd } n; \end{cases}$

2. $G_k(\mathbb{H}P^n) \supseteq (24, n+2)\gamma_n\nu^+_{4n+3} \circ \pi_k(\mathbb{S}^{4n+6})$. *In particular, we derive* $G_{4n+6}(\mathbb{H}P^n) \supseteq (24, n+2)\gamma_{n*}\pi_{4n+6}(\mathbb{S}^{4n+3}) \cong \mathbb{Z}_{\frac{24}{(24,n+2)}}$ *for* $n \geq 2$.

Observe that Theorem 2.49 improves [40, Corollary 2.7]. Namely, we obtain:

Corollary 2.50. 1. $G'_{8n+9}(\mathbb{H}P^{2n}) = 0$ *and* $G'_{8n+10}(\mathbb{H}P^{2n+1}) = \gamma_{2n+1*}\pi_{8n+10}(\mathbb{S}^{8n+7})$ *for* $n \not\equiv 0 \pmod 3$.

2. $G'_{4n+14}(\mathbb{H}P^n) = \gamma_{n*}\pi_{4n+14}(\mathbb{S}^{4n+3})$ *for* $n \equiv 5, 9 \pmod{12}$ *with* $n \geq 5$ *and* $n \equiv 15, 23 \pmod{24}$.

Proof. (1) is a direct consequence of Theorem 2.25(4) and Theorem 2.49(2).
We know from [56] and [57] that

$$\pi_{4n+13}(Sp(n)) \cong \begin{cases} \mathbb{Z}_8 & \text{for } n \equiv 1 \pmod 4 \text{ and } n \geq 5, \\ \mathbb{Z}_{(128,4(n-3))} & \text{for } n \equiv 3 \pmod 4. \end{cases}$$

So, for n satisfying $(24, n+2) = 1, n \equiv 1 \pmod 4$, or $n \equiv 7 \pmod 8$, in view of (1.138), (2.9), and (2.78), we obtain

$$\begin{aligned}\Delta_{\mathbb{H}}\zeta_{4n+3} = (\Delta_{\mathbb{H}}\iota_{4n+3}) \circ \zeta_{4n+2} &\in \{\omega_n, \nu_{4n+2}, \sigma_{4n+5}\} \circ 16\iota_{4n+13} \\ &\subseteq \pi_{4n+13}(Sp(n)) \circ 16\iota_{4n+13} = 0.\end{aligned}$$

Because $\pi_{4n+13}(Sp(n); 2) = \pi_{4n+13}(Sp(n))$, Lemma 2.33 leads to (2) and the proof is established. □

By the use of the exact sequence $(\mathcal{SH}^2_{11})$, Lemma 2.48(2), and the fact that $\pi_{12}(Sp(2)) = \{i_{2,\mathbb{H}}\varepsilon_3\} \cong \mathbb{Z}_2$ [54, Theorem 5.1], we obtain

$$\Delta_{\mathbb{H}}\eta_{11} = i_{2,\mathbb{H}}\varepsilon_3.$$

Then, by the relation (1.71)(2), we get:

Example 2.51. $G_{18+k}(\mathbb{H}P^2) \supseteq \{\gamma_2\eta^k_{11}\sigma_{11+k}\} \cong \mathbb{Z}_2$ for $k = 1, 2$.

The result [33, (2.10)(b)] implies

$$p_{n*}\omega_n = \pm(n+1)\nu^+_{4n-1}, \text{ where } p_n = p_{n,\mathbb{H}} \text{ and } n \geq 2. \tag{2.79}$$

Then, we obtain:

Lemma 2.52. *The elements $\eta_{16n-1}\bar{\kappa}_{16n}$ are $\mathbb{H}$-cyclic for $n \geq 1$.*

Proof. By (2.24) and (2.25), it holds

$$\{\nu_n^2, 2\iota_{n+6}, \kappa_{n+6}\} \ni \eta_n\bar{\kappa}_{n+1} \pmod{\sigma_n\kappa_{n+7}} \text{ for } n \geq 9.$$

By Lemma 2.48(3), $\sharp(\omega_{4n}\nu^+_{16n+2}) = 2(12, 2n+1)$. So, we can define the Toda bracket

$$\{\omega_{4n}\nu^+_{16n+2}, 2(12, 2n+1)\iota_{16n+5}, \kappa_{16n+5}\} \subseteq \pi_{16n+20}(Sp(4n)).$$

In view of (2.24) and the fact that $\pi^{16n+5}_{16n+20} = \{\rho_{16n+5}, \eta_{16n+5}\kappa_{16n+6}\}$, we have $(\nu^2_{16n-1})^+ \circ \pi^{16n+5}_{16n+20} = 0$. Hence, by (2.79), we obtain

$$p_{4n*}\{\omega_{4n}\nu^+_{16n+2}, 2(12, 2n+1)\iota_{16n+5}, \kappa_{16n+5}\} \subseteq \{(\nu^2_{16n-1})^+, 2(12, 2n+1)$$
$$\iota_{16n+5}, \kappa_{16n+5}\} \ni \eta_{16n-1}\bar{\kappa}_{16n} \pmod{\sigma_{16n-1}\kappa_{16n+6}}.$$

Then, the relation $\sigma_{16n-1}\kappa_{16n+6} = 0$ for $n \geq 1$ [49, Proposition 7.2] yields

$$p_{4n*}\{\omega_{4n}\nu^+_{16n+2}, 2(12, 2n+1)\iota_{16n+5}, \kappa_{16n+5}\} = \eta_{16n-1}\bar{\kappa}_{16n}.$$

This completes the proof. □

We show:

Corollary 2.53. $G'_{16n+k}(\mathbb{H}P^{4n-1}) = \gamma_{4n-1*}\pi_{16n+k}(\mathbb{S}^{16n-1})$ *for $k = 20$, $n \geq 1$ and $k = 21$, $n \geq 2$.*

Proof. By Corollary 2.2, (2.29), and (2.78), the element $E^2(\lambda\nu_{31}) = \nu_{15}\nu^*_{18} + \nu_{15}\xi_{18}$ is $\mathbb{H}$-cyclic. Further, (2.33) and (2.78) yield $\Delta_{\mathbb{H}}\sigma^3_{16n-1} = 0$ and Lemma 2.52 implies $\Delta_{\mathbb{H}}(\eta_{16n-1}\bar{\kappa}_{16n}) = 0$.

By means of (2.78), we get that $\nu_{8n-1}\bar{\sigma}_{8n+2}$ is $\mathbb{H}$-cyclic. In view of (2.9), (2.28), and Lemma 2.52, it holds $\Delta_{\mathbb{H}}(\eta^2_{16n-1}\bar{\kappa}_{16n+1}) = \Delta_{\mathbb{H}}(\eta_{16n-1}\bar{\kappa}_{16n})\eta_{16n+19} = 0$.

This, Lemma 2.33, and the group structures of $\pi_{16n+k}(\mathbb{S}^{16n-1})$ for $k = 20, n \geq 1$ and $k = 21, n \geq 2$ [49, Theorems A and B] complete the proof. □

Finally, we show:

Proposition 2.54. 1. $G_{18}(\mathbb{H}P^2) \supseteq 40\gamma_{2*}\pi_{18}(\mathbb{S}^{11})$;
2. $G_{21}(\mathbb{H}P^2) \supseteq 2\gamma_{2*}\pi_{21}(\mathbb{S}^{11})$;
3. $G_{22}(\mathbb{H}P^2) \supseteq 8\gamma_{2*}\pi_{22}(\mathbb{S}^{11}) \cong \mathbb{Z}_{63}$;
4. $G_{22}(\mathbb{H}P^3) \supseteq 4\gamma_{3*}\pi_{22}(\mathbb{S}^{15}) \cong \mathbb{Z}_{60}$.

Proof. By [54, Theorem 5.1] and [55, Lemmas 3.1–4], we see that:

1. $\mathrm{Ker}\{\Delta_{\mathbb{H}} : \pi_{18}(\mathbb{S}^{11}) \to \pi_{17}(Sp(2))\} = 40\pi_{18}(\mathbb{S}^{11})$;
2. $\mathrm{Ker}\{\Delta_{\mathbb{H}} : \pi_{21}(\mathbb{S}^{11}) \to \pi_{20}(Sp(2))\} = 2\pi_{21}(\mathbb{S}^{11})$;
3. $\mathrm{Ker}\{\Delta_{\mathbb{H}} : \pi_{22}(\mathbb{S}^{11}) \to \pi_{21}(Sp(2))\} = 8\pi_{22}(\mathbb{S}^{11})$;
4. $\mathrm{Ker}\{\Delta_{\mathbb{H}} : \pi_{22}(\mathbb{S}^{15}) \to \pi_{21}(Sp(3))\} = 4\pi_{22}(\mathbb{S}^{15})$.

Hence, by Lemma 2.33(1), we obtain the assertion and this completes the proof. □

We close this section with:

Question 2.55. What about $G_k(\mathbb{H}P^n) \cap i_{\mathbb{H}*}E\pi_{k-1}(\mathbb{S}^3)$?

2.8 The Case of the Cayley Projective Plane

Let the homotopy cofiber $\mathbb{K}P^2 = \mathbb{S}^8 \cup_{\sigma_8^+} e^{16}$ of the map $\sigma_8^+ : \mathbb{S}^{15} \to \mathbb{S}^8$ be the Cayley projective plane and write $i_{\mathbb{K}} : \mathbb{S}^8 \hookrightarrow \mathbb{K}P^2$ for the inclusion map. By [23, Theorem 1], we obtain

$$G_8(\mathbb{K}P^2) = 0$$

proved in [19] and [40] as well.

We recall the fibration

$$p : F_4 \longrightarrow F_4/Spin(9) = \mathbb{K}P^2$$

and the induced exact sequence:

$$\cdots \longrightarrow \pi_n(\mathbb{K}P^2) \xrightarrow{\Delta_{\mathbb{K}}} \pi_{n-1}(Spin(9)) \xrightarrow{i_*} \pi_{n-1}(F_4) \xrightarrow{p_*} \cdots . \qquad (2.80)$$

Since $\pi_7(F_4) = 0$, $\pi_7(Spin(9)) \cong \mathbb{Z}$ and $\pi_8(\mathbb{K}P^2) \cong \mathbb{Z}$ [51, p. 132, Table], it follows that

$$\Delta_{\mathbb{K}} : \pi_8(\mathbb{K}P^2) \longrightarrow \pi_7(Spin(9))$$

is an isomorphism.

Identifying $\pi_n(Spin(9))$ with $\pi_n(SO(9))$ for $n \geq 2$, we may set $\Delta_{\mathbb{K}}(i_{\mathbb{K}}) = \pm[\iota_7]_9$. Then, in view of (2.9), we have $\Delta_{\mathbb{K}}(i_{\mathbb{K}}E\alpha) = \pm[\iota_7]_9\alpha$ for $\alpha \in \pi_{n-1}(\mathbb{S}^7)$ which implies:

Lemma 2.56. *If $\pi_k(Spin(9))$ is a 2-primary group and p is an odd prime then $i_{\mathbb{K}}E\pi_k(\mathbb{S}^7; p) \subseteq G_{k+1}(\mathbb{K}P^2)$.*

The groups $\pi_n(\mathbb{K}P^2)$ have been determined in [51] for $n \le 23$ and $\pi_n(\mathbb{K}P^2;2)$ for $n \le 38$ in [29]. Let

$$Y = \mathbb{S}^7 \cup_{\sigma'\sigma_{14}+\alpha'} e^{22}$$

be the homotopy cofiber of the map $\sigma'\sigma_{14} + \alpha' : \mathbb{S}^{21} \to \mathbb{S}^7$, where $E\alpha' = [[\iota_8, \iota_8], \iota_8] \neq 0$ (1.4). By [51, Proposition 7.1], the 2- and 3-primary components of $\pi_n(\mathbb{K}P^2)$ are obtained by determining $\pi_{n-1}(Y)$ for $n \le 28$, and the cell complex Y is the 28-skeleton of $\Omega(\mathbb{K}P^2)$.

Recall from [85, Chap. I] that a map $f : EX_1 \to X_2$ determines $\Omega_0 f : X_1 \to \Omega X_2$ defined by $\Omega_0 f = (\Omega f) \circ i$, where $i : X_1 \to \Omega(EX_1)$ is the adjoint map of id_{EX_1}. Thus, we get the factorization

$$\begin{array}{ccc} \mathbb{S}^7 & \xrightarrow{\Omega_0 i_{\mathbb{K}}} & \Omega(\mathbb{K}P^2) \\ {\scriptstyle i_Y}\big\downarrow & \nearrow & \\ Y. & & \end{array} \tag{2.81}$$

Further, the associated with Y sequence

$$\cdots \to \pi_k(\mathbb{S}^7) \xrightarrow{i_{Y*}} \pi_k(Y) \xrightarrow{p_{Y*}} \pi_k(\mathbb{S}^{22}) \xrightarrow{\Delta} \pi_{k-1}(\mathbb{S}^7) \to \cdots \tag{2.82}$$

is exact for $k \le 27$ [51, (7.6)], where $i_Y : \mathbb{S}^7 \to Y$ and $p_Y : Y \to \mathbb{S}^{22}$ are the inclusion and quotient maps, respectively. We need:

Lemma 2.57. 1. $\pi_n(\mathbb{K}P^2) = i_{\mathbb{K}*}E\pi_{n-1}(\mathbb{S}^7) \cong \pi_{n-1}(\mathbb{S}^7)$ *for* $n \le 21$ *or* $n = 25, 27, 28$*;*
2. $\pi_{22}(\mathbb{K}P^2) = i_{\mathbb{K}*}\{\kappa_8\} \cong \mathbb{Z}_4$ *and* $i_{\mathbb{K}}(E\sigma')\sigma_{15} = 0$*;*
3. $\pi_{23}(\mathbb{K}P^2) = \{\widetilde{8\iota_{22}}\} \oplus i_{\mathbb{K}*}\{E\rho''^{+}, \bar{\varepsilon}_8, (E\sigma')\varepsilon_{15}\} \cong \mathbb{Z} \oplus \mathbb{Z}_{120} \oplus (\mathbb{Z}_2)^2$, *where* $\widetilde{8\iota_{22}} \in \{i_{\mathbb{K}}, (E\sigma')\sigma_{15}, 8\iota_{22}\}_1$ *and* $i_{\mathbb{K}}(E\sigma')\varepsilon_{15} = i_{\mathbb{K}}(E\sigma')\bar{\nu}_{15}$*;*
4. $\pi_{24}(\mathbb{K}P^2) = i_{\mathbb{K}*}\{(E\sigma')\mu_{15}, \mu_8\sigma_{17}, \eta_8\bar{\varepsilon}_9\} \cong (\mathbb{Z}_2)^3$ *and* $i_{\mathbb{K}}\eta_8\bar{\varepsilon}_9 = i_{\mathbb{K}}(E\sigma')\nu_{15}^3 = i_{\mathbb{K}}(E\sigma')(\eta_{15}\varepsilon_{16} + \sigma_{15}\eta_{22}^2) = i_{\mathbb{K}}(E\sigma')\eta_{15}\varepsilon_{16}$*;*
5. *there is the short exact sequence*

$$0 \to \mathbb{Z}_{24} \oplus \mathbb{Z}_2 \to \pi_{26}(\mathbb{K}P^2) \to \mathbb{Z}_{24} \to 0$$

and $\pi_{26}(\mathbb{K}P^2;2) = \{\eta_8\bar{\mu}_9, \tilde{\nu}_{22}\} \cong \mathbb{Z}_2 \oplus \mathbb{Z}_{64}$ *for* $\tilde{\nu}_{22} \in \{i_{\mathbb{K}}, (E\sigma')\sigma_{15}, \nu_{22}\}$ *with* $8\tilde{\nu}_{22} = \pm i_{\mathbb{K}}\zeta_8\sigma_{19}$*;*
6. $\pi_{29}(\mathbb{K}P^2;2) = \{\tilde{\nu}_{22}\nu_{26}, i_{\mathbb{K}}\eta_8\bar{\kappa}_9, i_{\mathbb{K}}(E\sigma')\kappa_{15}\} \cong \mathbb{Z}_2 \oplus (\mathbb{Z}_2)^2$.

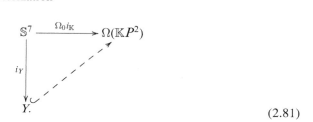

Proof. First, the homotopy exact sequence

$$\cdots \to \pi_n(\mathbb{S}^8) \stackrel{i_{\mathbb{K}*}}{\to} \pi_n(\mathbb{K}P^2) \to \pi_n(\mathbb{K}P^2, \mathbb{S}^8) \stackrel{\partial}{\to} \pi_{n-1}(\mathbb{S}^8) \to \cdots$$

of the pair $(\mathbb{K}P^2, \mathbb{S}^8)$ and (1.27)(3) implies $i_{\mathbb{K}*}E\pi_{n-1}(\mathbb{S}^7) \subseteq \pi_n(\mathbb{K}P^2)$. Blakers–Massey theorem leads to isomorphism $\pi_n(\mathbb{K}P^2, \mathbb{S}^8) \stackrel{\cong}{\to} \pi_n(\mathbb{S}^{16})$ for $n \leq 23$. Hence, those yield $\pi_n(\mathbb{K}P^2) = i_{\mathbb{K}*}E\pi_{n-1}(\mathbb{S}^7) \cong \pi_{n-1}(\mathbb{S}^7)$ for $n \leq 21$.

Next, we make use of (2.82) to describe the groups $\pi_n(\mathbb{K}P^2)$ for $n = 23, 24, 26$. By Blakers–Massey theorem, $\pi_{22}(Y, \mathbb{S}^7) \cong \mathbb{Z}$ and $\pi_{23}(Y, \mathbb{S}^7) \cong \pi_{23}(\mathbb{S}^{22})$. Since $(\sigma'\sigma_{14} + \alpha')\eta_{21} = \sigma'(\bar{\nu}_{14} + \varepsilon_{14})$, we obtain the group $\pi_{22}(Y)$. Further, in view of (2.82), we have $\Delta(\iota_{22})_{(2)} = \sigma'\sigma_{14}$. Because id_Y is regarded as an extension of the inclusion map $i_Y : \mathbb{S}^7 \to Y$, [85, Proposition 1.7] yields a coextension $-\widetilde{8\iota_{21}} \in \{i_Y, \sigma'\sigma_{14}, 8\iota_{21}\}$. On the other hand, in view of the factorization (2.81), the result [85, Proposition 1.3] leads to the formula

$$-\Omega_0\{i_{\mathbb{K}}, (E\sigma')\sigma_{15}, 8\iota_{22}\}_1 = \{i_Y, \sigma'\sigma_{14}, 8\iota_{21}\}$$

which yields the corresponding coextension $\widetilde{8\iota_{22}} \in \{i_{\mathbb{K}}, (E\sigma')\sigma_{15}, 8\iota_{22}\}_1$ generating the direct summand $\mathbb{Z}$ of the group $\pi_{23}(\mathbb{K}P^2)$.

By the parallel argument, we have the group $\pi_{23}(Y)$ and the assertions for $n = 23, 24$ follow.

We know that $\pi_{26}(\mathbb{S}^{22}) = 0$, $\pi_{25}(\mathbb{S}^7) \cong \mathbb{Z}_{24} \oplus \mathbb{Z}_2$, $\pi_{25}(\mathbb{S}^{22}) \cong \mathbb{Z}_{24}$, and the relation $(\sigma'\sigma_{14} + \alpha')\nu_{21} = 0$ (1.85). So, the short exact sequence (2.82) leads to the short exact one

$$0 \to \mathbb{Z}_{24} \oplus \mathbb{Z}_2 \to \pi_{26}(\mathbb{K}P^2) \to \mathbb{Z}_{24} \to 0$$

and, as above, to a coextension $\tilde{\nu}_{22} \in \{i_{\mathbb{K}}, (E\sigma')\sigma_{15}, \nu_{22}\}$. In view of (1.65),

$$8\{i_{\mathbb{K}}, (E\sigma')\sigma_{15}, \nu_{22}\} = -i_{\mathbb{K}*}\{(E\sigma')\sigma_{15}, \nu_{22}, 8\iota_{25}\} \supseteq -i_{\mathbb{K}}(E\sigma')\{\sigma_{15}, \nu_{22}, 8\iota_{25}\} \ni$$

$-i_{\mathbb{K}}(E\sigma')\zeta_{15} = x i_{\mathbb{K}}\zeta_8\sigma_{19}$ (mod $8i_{\mathbb{K}*}\pi^8_{26} = 0$) for x as in (1.65). Consequently, we derive that $8\tilde{\nu}_{22} = \pm i_{\mathbb{K}}\zeta_8\sigma_{19}$ and (5) follows.

(6) is determined by [29, Theorem 5.3] and the proof is complete. □

Now, we show:

Lemma 2.58. 1. $\mathrm{Ker}\{E^8 : \pi_{n-1}(\mathbb{S}^7) \to \pi_{n+7}(\mathbb{S}^{15})\} \subseteq (i_{\mathbb{K}}E)_*^{-1}P_n(\mathbb{K}P^2)$. *In particular,* $i_{\mathbb{K}}E(\sigma'\alpha) \in P_n(\mathbb{K}P^2)$ *provided that* $\sharp\alpha = 2$ *for* $\alpha \in \pi^{14}_{n-1}$;

2. *If* $\sharp(i_{\mathbb{K}}(E\sigma')(E^8\alpha)) = 2^m$ *with* $m > 0$ *for* $\alpha \in \pi^7_{n-1}$ *then* $2^k i_{\mathbb{K}}(E\alpha) \notin P_n(\mathbb{K}P^2)$ *provided that* $k < m$.

Proof. (1) is obtained from the formula parallel to (2.59). (2) follows from the formula

$$\pm [i_{\mathbb{K}}(E\alpha), i_{\mathbb{K}}] = i_{\mathbb{K}}(E\sigma')(E^8\alpha) \tag{2.83}$$

determined by Lemma 1.2(2) and (1.22) for $\alpha \in \pi_{n-1}(\mathbb{S}^7)$. □

Hence, we deduce:

Corollary 2.59.

$$P_n(\mathbb{K}P^2) \ni \begin{cases} i_{\mathbb{K}}E(\sigma'\eta_{14}) & \textit{for} \quad n = 16, \\ i_{\mathbb{K}}E(\sigma'\eta_{14}^2) & \textit{for} \quad n = 17, \\ i_{\mathbb{K}}E(\sigma'\nu_{14}) = x i_{\mathbb{K}}E(\nu_7\sigma_{10}) & \textit{for an odd integer } x \textit{ and} \quad n = 18, \\ i_{\mathbb{K}}E(\bar{\nu}_7\nu_{15}) & \textit{for} \quad n = 19, \\ i_{\mathbb{K}}E(\nu_7\sigma_{10}\nu_{17}) & \textit{for} \quad n = 21, \\ i_{\mathbb{K}}E(\bar{\nu}_7\nu_{15}^2) & \textit{for} \quad n = 22, \\ i_{\mathbb{K}}E(\sigma'\varepsilon_{14}) = i_{\mathbb{K}}E(\sigma'\bar{\nu}_{14}) & \textit{for} \quad n = 23, \\ i_{\mathbb{K}}E(\sigma'\mu_{14}), i_{\mathbb{K}}E(\eta_7\bar{\varepsilon}_8) & \textit{for} \quad n = 24, \\ i_{\mathbb{K}}E(\sigma'\eta_{14}\mu_{15}) & \textit{for} \quad n = 25. \end{cases}$$

Proof. Because, by Lemma 2.57(3), $i_{\mathbb{K}}E(\sigma'\varepsilon_{14}) = i_{\mathbb{K}}E(\sigma'\bar{\nu}_{14})$ and $E^2\sigma' = 2\sigma_9$ (1.22), Lemma 2.58(1) leads to:

$$P_n(\mathbb{K}P^2) \ni \begin{cases} i_{\mathbb{K}}E(\sigma'\eta_{14}) & \text{for} \quad n = 16, \\ i_{\mathbb{K}}E(\sigma'\eta_{14}^2) & \text{for} \quad n = 17, \\ i_{\mathbb{K}}E(\sigma'\varepsilon_{14}) = i_{\mathbb{K}}E(\sigma'\bar{\nu}_{14}) & \text{for} \quad n = 23, \\ i_{\mathbb{K}}E(\sigma'\mu_{14}) & \text{for} \quad n = 24, \\ i_{\mathbb{K}}E(\sigma'\eta_{14}\mu_{15}) & \text{for} \quad n = 25. \end{cases}$$

By means of (1.82), $i_{\mathbb{K}}E(\sigma'\nu_{14}) = x i_{\mathbb{K}}E(\nu_7\sigma_{10})$ for an odd integer x. Hence, (1.84) and Lemma 2.58(1) lead to $i_{\mathbb{K}}E(\sigma'\nu_{14}) = x i_{\mathbb{K}}E(\nu_7\sigma_{10}) \in P_{18}(\mathbb{K}P^2)$. Next, Lemma 2.58(1) and the relation (1.129) yield $i_{\mathbb{K}}E(\bar{\nu}_7\nu_{15}) \in P_{19}(\mathbb{K}P^2)$. Because $i_{\mathbb{K}}E(\nu_7\sigma_{10}) \in P_{18}(\mathbb{K}P^2)$, in view of Corollary 2.2, the element $i_{\mathbb{K}}E(\nu_7\sigma_{10}\nu_{17}) \in P_{21}(\mathbb{K}P^2)$. In virtue of (1.84) and (1.107)(1), Lemma 2.58(1) leads to $i_{\mathbb{K}}E(\eta_7\bar{\varepsilon}_8) \in P_{24}(\mathbb{K}P^2)$ and the proof is complete. □

Next, we show:

Theorem 2.60. 1. $P_n(\mathbb{K}P^2) = 0$ *for* $n = 9, 10, 12, 13, 14, 20$;
2. $G_{11}(\mathbb{K}P^2) = P_{11}(\mathbb{K}P^2) = 8\pi_{11}(\mathbb{K}P^2) \cong \mathbb{Z}_3$;
3. $8\pi_{15}(\mathbb{K}P^2) \subseteq G_{15}(\mathbb{K}P^2)$;
4. $i_{\mathbb{K}*}E\{\sigma'\eta_{14}\} \subseteq P_{16}(\mathbb{K}P^2) \subseteq i_{\mathbb{K}*}E\{\sigma'\eta_{14}, \eta_7\sigma_8\}$;
5. $i_{\mathbb{K}*}E\{\sigma'\eta_{14}^2\} \subseteq P_{17}(\mathbb{K}P^2) \subseteq i_{\mathbb{K}*}E\{\sigma'\eta_{14}^2, \nu_7^3 + \eta_7\varepsilon_8\} = i_{\mathbb{K}*}E\{\sigma'\eta_{14}, \eta_7\sigma_8\} \circ \eta_{16}$;

6. $8\pi_{18}(\mathbb{K}P^2) \subseteq G_{18}(\mathbb{K}P^2) \subseteq P_{18}(\mathbb{K}P^2) = i_{\mathbb{K}*}\{\nu_8\sigma_{11}, \beta_1(8)\} \cong \mathbb{Z}_{24}$;
7. $P_{19}(\mathbb{K}P^2; 2) = i_{\mathbb{K}*}E\{\bar{\nu}_7\nu_{15}\} \cong \mathbb{Z}_2$;
8. $2\pi_{21}(\mathbb{K}P^2) \subseteq G_{21}(\mathbb{K}P^2) \subseteq P_{21}(\mathbb{K}P^2) = \pi_{21}(\mathbb{K}P^2)$.
9. $P_{22}(\mathbb{K}P^2) = 2\pi_{22}(\mathbb{K}P^2)$.

Proof. The results about the lower bounds of $P_n(\mathbb{K}P^2)$ are obtained in Lemma 2.58.

Since, by (2.75) and Lemma 2.57(1), $\pi_n(\mathbb{K}P^2) \cong \pi_{n-1}(\mathbb{S}^7) = 0$ for $n = 12, 13, 20$, the assertion for $n = 12, 13, 20$ of (1) follows.

By Lemma 2.57(1),

$$i_{\mathbb{K}}(E\sigma')\eta_{15} \neq 0,\ i_{\mathbb{K}}(E\sigma')\eta_{15}^2 \neq 0,$$

$$i_{\mathbb{K}}(E\sigma')\nu_{15}^2 = i_{\mathbb{K}}\nu_8\sigma_{11}\nu_{18} \neq 0.$$

This and Lemma 2.58(2) lead to the assertion for $n = 9, 10, 14$ of (1).

Then, by the facts that $\pi_{11}(\mathbb{K}P^2) = i_{\mathbb{K}*}\{\nu_8^+\} \cong \mathbb{Z}_{24}$ Lemma 2.57(1), $\pi_{10}(Spin(9)) \cong \mathbb{Z}_8$ [51, p. 132, Table], Lemma 2.56 implies

$$i_{\mathbb{K}}\alpha_1(8) \in G_{11}(\mathbb{K}P^2). \tag{2.84}$$

Moreover, $\sharp((i_{\mathbb{K}}(E\sigma')\nu_{18}) = 8$ (1.82) and Lemma 2.57(1). This and Lemma 2.58(2) yield $P_{11}(\mathbb{K}P^2; 2) = 0$, and (2) follows.

Because $\pi_{14}(\mathbb{S}^7) = \{\sigma'^+\} \cong \mathbb{Z}_{120}$ (2.75), we get $\pi_{15}(\mathbb{K}P^2) = \{i_{\mathbb{K}}(E\sigma'^+)\} \cong \mathbb{Z}_{120}$. Then, in view of Lemma 2.56 and Lemma 2.57(1), $8\pi_{15}(\mathbb{K}P^2) \subseteq G_{15}(\mathbb{K}P^2)$ which yields (3).

We recall from (2.75) that $\pi_{15}(\mathbb{S}^7) = \{\sigma'\eta_{14}, \bar{\nu}_7, \varepsilon_7\} \cong (\mathbb{Z}_2)^3$. By Lemma 2.57(1),

$$\pi_{16}(\mathbb{K}P^2) = i_{\mathbb{K}*}E\{\sigma'\eta_{14}, \bar{\nu}_7, \varepsilon_7\}.$$

Next, by Lemma 2.57(3), $i_{\mathbb{K}}(E\sigma')\bar{\nu}_{15} = i_{\mathbb{K}}(E\sigma')\varepsilon_{15} \neq 0$ and (1.69)(1) yields $i_{\mathbb{K}}(E\sigma')(\bar{\nu}_{15} + \varepsilon_{15}) = i_{\mathbb{K}}(E\sigma')(\eta_{15}\sigma_{16}) = 0$. This leads to (4).

We recall from (2.75) that

$$\pi_{16}(\mathbb{S}^7) = \{\sigma'\eta_{14}^2, \nu_7^3, \mu_7, \eta_7\varepsilon_8\} \cong (\mathbb{Z}_2)^4.$$

By Lemma 2.57(4), $i_{\mathbb{K}}(E\sigma')\nu_{15}^3 = i_{\mathbb{K}}\eta_8\bar{\varepsilon}_9 \neq 0$, $i_{\mathbb{K}}(E\sigma')\mu_{15} \neq 0$.

Then, by Lemma 2.58(2), $i_{\mathbb{K}}\nu_8^3 \notin P_{17}(\mathbb{K}P^2)$ and $i_{\mathbb{K}}\mu_8 \notin P_{17}(\mathbb{K}P^2)$. Moreover, by (1.69)(1) and (1.72), we obtain $i_{\mathbb{K}}(E\sigma')(\eta_{15}\varepsilon_{16}) = i_{\mathbb{K}}(E\sigma')(\nu_{15}^3) \neq 0$ and $i_{\mathbb{K}}(E\sigma')(\nu_{15}^3 + \eta_{15}\varepsilon_{16}) = i_{\mathbb{K}}(E\sigma')(\sigma_{15}\eta_{22}^2) = 0$. This leads to (5).

We recall from (2.75) that $\pi_{17}(\mathbb{S}^7) = \{\nu_7\sigma_{10}, \eta_7\mu_8, \beta_1(7)\} \cong \mathbb{Z}_{24} \oplus \mathbb{Z}_2$. By Lemma 2.57(1),

$$\pi_{18}(\mathbb{K}P^2) = i_{\mathbb{K}*}E\{\nu_7\sigma_{10}, \eta_7\mu_8, \beta_1(7)\} \cong \mathbb{Z}_{24} \oplus \mathbb{Z}_2.$$

Since $\pi_{17}(Spin(9)) \cong \mathbb{Z}_8 \oplus (\mathbb{Z}_2)^2$ [51, p. 132, Table], Lemma 2.56 implies $i_{\mathbb{K}}\beta_1(8) \in G_{18}(\mathbb{K}P^2)$. In view of Corollary 2.59, the element $i_{\mathbb{K}}E(\nu_7\sigma_{10}) \in P_{18}(\mathbb{K}P^2)$. By Lemma 2.57(1) and the fact from (2.75) that

$$\pi_{24}(\mathbb{S}^7) = \{\sigma'\eta_{14}\mu_{15}, \nu_7\kappa_{10}, \bar{\mu}_7, \eta_7\mu_8\sigma_{17}\} \cong (\mathbb{Z}_2)^4,$$

we see that $i_{\mathbb{K}}E(\sigma'\eta_{14}\mu_{15}) \neq 0$ and, in view of Lemma 2.58(2), $i_{\mathbb{K}}E(\eta_7\mu_8) \notin P_{18}(\mathbb{K}P^2)$. This leads to (6).

Next, we recall from (2.75) that

$$\pi^7_{18} = \{\zeta_7, \bar{\nu}_7\nu_{15}\} \cong \mathbb{Z}_8 \oplus \mathbb{Z}_2.$$

But, $i_{\mathbb{K}}(E\sigma')\zeta_{15} = xi_{\mathbb{K}}\zeta_8\sigma_{19}$ for x as in (1.65), and $\sharp(i_{\mathbb{K}}(\zeta_8\sigma_{19})) = 8$ (Lemma 2.57(5)), so Lemma 2.58(2) yields $2^k i_{\mathbb{K}}(E\zeta_7) \notin P_{19}(\mathbb{K}P^2)$ with $k < 3$. In view of Corollary 2.59, the element $i_{\mathbb{K}}E(\bar{\nu}_7\nu_{15}) \in P_{19}(\mathbb{K}P^2)$ and this leads to (7).

By Lemma 2.57(1) and the fact from (2.75) that $\pi_{20}(\mathbb{S}^7) = \{\nu_7\sigma_{10}\nu_{17}, \alpha_1(7)\beta_1(10)\} \cong \mathbb{Z}_6$, we obtain $\pi_{21}(\mathbb{K}P^2) = i_{\mathbb{K}*}\{\nu_8\sigma_{11}\nu_{18}, \alpha_1(8)\beta_1(11)\}$. In view of (2.84), we get

$$i_{\mathbb{K}}\alpha_1(8)\beta_1(11) \in G_{21}(\mathbb{K}P^2)$$

and, by means of Corollary 2.59, the element $i_{\mathbb{K}}E(\nu_7\sigma_{10}\nu_{17}) \in P_{21}(\mathbb{K}P^2)$. Those lead to (8).

Recall $\pi_{22}(\mathbb{K}P^2) = \{i_{\mathbb{K}}\kappa_8\} \cong \mathbb{Z}_4$ Lemma 2.57(2). By Lemma 2.57(6) and (2.83), $[i_{\mathbb{K}}\kappa_8, i_{\mathbb{K}}] = i_{\mathbb{K}}(E\sigma')\kappa_{15} \neq 0$. Hence, in view of (1.109) and Corollary 2.59, (9) follows and the proof is complete. □

We show:

Proposition 2.61. *If* $\sharp(\sigma_9 E^9\alpha) = m > 1$ *for* $\alpha \in \pi^7_{n-1}$ *then* $k(i_{\mathbb{K}}E\alpha) \notin \mathrm{Ker}\{\Delta_{\mathbb{K}} : \pi_n(\mathbb{K}P^2) \to \pi_{n-1}(Spin(9))\}$ *unless* $k \equiv 0 \pmod m$.

Proof. Suppose that $\Delta_{\mathbb{K}}(ki_{\mathbb{K}}E\alpha) = 0$ with $k \not\equiv 0 \pmod m$ for $\sharp(\sigma_9 E^9\alpha) = m$. Then, $0 = kJ\Delta_{\mathbb{K}}(i_{\mathbb{K}}E\alpha) = \pm kJ([\iota_7]_9\alpha) = \pm k\sigma_9 E^9\alpha$ leads to a contradiction.

□

Using (2.75), we obtain:

Example 2.62. The following $\alpha \in \pi^7_{n-1}$ subject:

1. $\sharp(\sigma_9 E^9\alpha) = 2$ for: $\alpha = \eta_7, \eta_7^2, \nu_7^2, \eta_7\sigma_8, \bar{\nu}_7, \varepsilon_7, \nu_7^3, \mu_7, \eta_7\varepsilon_8, \eta_7\mu_8\ \nu_7\kappa_{10}, \bar{\mu}_7, \mu_7\sigma_{16}$;
2. $\sharp(\sigma_9 E^9\alpha) = 8$ for: $\alpha = \nu_7, \sigma', \zeta_7$.

Owing to Example 2.62 and Theorem 2.60, we propose:

Question 2.63. $G_n(\mathbb{K}P^2; 2) = 0$?

Chapter 3
Gottlieb and Whitehead Center Groups of Moore Spaces

This chapter takes up the systematic study of the Gottlieb groups $G_{n+k}(M(A,n))$ of Moore spaces $M(A,n)$ for an abelian group A and $n \geq 2$. The groups $G_{n+k}(M(A,n))$ and $G_{n+k}(M(A \oplus \mathbb{Z},n))$ are determined for $k = 0,1,2,3,4,5$ and $n \geq 2$ provided A is finite.

3.1 Gottlieb and Whitehead Center Groups of Mod 2 Moore Spaces

Given $\alpha \in \pi_n(EX)$, consider the generalized Whitehead product $[\iota_{EX},\alpha] \in [E^nX, EX]$ studied in [3] and define

$$P : \pi_n(EX) \longrightarrow [E^nX, EX]$$

by $P(\alpha) = [\iota_{EX},\alpha]$ for $\alpha \in \pi_n(EX)$.

Proposition 3.1. *If EX is the suspension of a space X then $G_n(EX) = \ker P$.*

Proof. For the inclusion maps $j_1 : EX \hookrightarrow EX \vee \mathbb{S}^n$ and $j_2 : \mathbb{S}^n \hookrightarrow EX \vee \mathbb{S}^n$, consider the generalized Whitehead product $[j_1, j_2] \in [E^nX, EX \vee \mathbb{S}^n]$. Then, $EX \times \mathbb{S}^n = (EX \vee \mathbb{S}^n) \cup_{[j_1,j_2]} C(E^nX)$, the mapping cone of $[j_1, j_2] : E^nX \to EX \vee \mathbb{S}^n$. Next, consider the diagram

© Springer International Publishing Switzerland 2014

M. Golasiński, J. Mukai, *Gottlieb and Whitehead Center Groups of Spheres, Projective and Moore Spaces*, DOI 10.1007/978-3-319-11517-7_3

$$\begin{array}{ccc} E^nX \xrightarrow{[j_1,j_2]} & EX\vee\mathbb{S}^n & \\ & \downarrow \qquad \searrow^{\iota_{EX}\vee\alpha} & \\ & EX\times\mathbb{S}^n & \quad EX. \end{array}$$

Then, $\alpha \in G_n(EX)$ if and only if there is an extension $EX\times\mathbb{S}^n \to EX$ of $\iota_{EX}\vee\alpha$. Equivalently, $(\iota_{EX}\vee\alpha)[j_1,j_2] = [\iota_{EX},\alpha] = 0$, i.e., $\alpha \in \ker P$. □

Let $\alpha : EA \to X$ and $\beta : EB \to X$. Then, in view of [3, Proposition 3.3], we have (the anti-commutativity)

$$[\alpha,\beta] = -(E\tau)^*[\beta,\alpha], \tag{3.1}$$

where $\tau : A\wedge B \to B\wedge A$ is the obvious twisting map.

Now, we recall generalizations of results stated in Lemma 1.2(1), (2), and (6). In virtue of [3, Proposition 5.12]: $[\alpha,\beta] = 0$ if and only if there is a map $m : EA\times EB \to X$ such that $m|_{EA} = \alpha$ and $m|_{EB} = \beta$. Consequently, $[\alpha,\beta] = 0$ and maps $\gamma : EA' \to EA$ and $\delta : EB' \to EB$ lead to $m' = m(\gamma\times\delta) : EA'\times EB' \to X$ with $m'|_{EA'} = \alpha\circ\gamma$ and $m'|_{EB'} = \beta\circ\delta$.

Hence, we may state

$$[\alpha,\beta] = 0 \quad \text{implies} \quad [\alpha\circ\gamma,\beta\circ\delta] = 0 \tag{3.2}$$

for any maps $\gamma : EA' \to EA$ and $\delta : EB' \to EB$.

In view of [11, Proposition 3.2], we have

$$[\alpha\circ E\gamma,\beta\circ E\delta] = [\alpha,\beta]\circ E(\gamma\wedge\delta) \tag{3.3}$$

for any maps $\gamma : A' \to A$ and $\delta : B' \to B$.

Further, by [3, Proposition 1.3], it holds

$$E[\alpha,\beta] = 0. \tag{3.4}$$

Given the inclusion maps $j_1 : EX \hookrightarrow EX\vee EY$ and $j_2 : EY \hookrightarrow EX\vee EY$ and identifying $EX\vee EY$ with $E(X\vee Y)$, we deduce:

Corollary 3.2 ([5, Proposition 2.3]). *Let $\alpha \in \pi_n(EX\vee EY)$. Then $\alpha \in G_n(EX\vee EY)$ if and only if $[j_1,\alpha] = [j_2,\alpha] = 0$.*

Proof. If $[\iota_{EX},\alpha] = 0$ for $\alpha \in \pi_n(EX)$ then (3.2) implies $[j_k,\alpha] = 0$ for $k = 1,2$.

Because the maps $j_1 : EX \hookrightarrow EX\vee EY$ and $j_2 : EY \hookrightarrow EX\vee EY$ determine $\iota_{EX\vee EY}$, we conclude that $[j_k,\alpha] = 0$ for $k = 1,2$ imply $[\iota_{EX\vee EY},\alpha] = 0$. □

Recall that $M^n = E^{n-2}\mathbb{R}P^2$ for $n \geq 3$ denotes the Moore space of type $(\mathbb{Z}_2, n-1)$ for the real projective plane $\mathbb{R}P^2$.

Next, we prove:

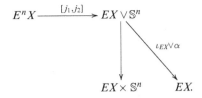

Proposition 3.3 (cf. [5, Sect. 4]). *If A is an abelian group and $M(A,n)$ is the Moore space of type (A,n) for $n \geq 2$ then $\mathrm{G}_n(M(A,n)) = 0$ if n is even or n is odd and A is torsion. In particular, $G_{n-1}(M^n) = 0$ for $n \geq 2$.*

Proof. Because the Hurewicz homomorphism

$$h_n : \pi_n(M(A,n)) \to H_n(M(A,n))$$

is an isomorphism, the result [22, Theorem 5-2] yields that $G_n(M(A,n)) = 0$ under the conditions stated for n and A.

For $M^2 = \mathbb{R}P^2$, the result Example 2.13 shows that $G_1(\mathbb{R}P^2) = 0$. If $n \geq 3$ then M^n is the Moore space of type $(\mathbb{Z}_2, n-1)$ and the above leads to $G_{n-1}(M^n) = 0$ to complete the proof. □

Remark 3.4. We have been informed by Jin-ho Lee that he has also shown that $G_{n-1}(M^n) = 0$ for $n \geq 3$ as a consequence of [22, Theorem 5-2].

If $A = \mathbb{Z}_m$ then $M(\mathbb{Z}_m, n) = \mathbb{S}^n \cup_{m\iota_n} e^{n+1}$ and write $i_{n+1} : \mathbb{S}^n \hookrightarrow M(\mathbb{Z}_m, n)$ for the canonical inclusion map. By means of [86, Theorem 4.4], it holds $\sharp\iota_{M(\mathbb{Z}_m,n)} = \begin{cases} 2m & \text{if } m \equiv 2 \pmod 4, \\ m & \text{if } m \not\equiv 2 \pmod 4. \end{cases}$ But, because $\pi_n(M(\mathbb{Z}_m,n)) = \{i_{n+1}\} \cong \mathbb{Z}_m$, we derive from Proposition 3.1 and 3.3:

Corollary 3.5. *The order $\sharp[i_{n+1}, \iota_{M(\mathbb{Z}_m,n)}] = m$.*

Next, we recall:

Theorem 3.6 (The Jacobi identity [78, 79, Theorem 1.1]). *Let A_1, A_2, and A_3 be CW-complexes such that EA_1, EA_2, and EA_3 are homotopy commutative H'-spaces and let $\alpha : EA_1 \to X$, $\beta : EA_2 \to X$, and $\gamma : EA_3 \to X$. Then in the group $[E(A_1 \wedge A_2 \wedge A_3), X]$ the generalized Whitehead product satisfies*

$$[[\alpha,\beta],\gamma] + (E\tau_{312})^*[[\gamma,\alpha],\beta] + (E\tau_{231})^*[[\beta,\gamma],\alpha] = 0,$$

where $\tau_{312} : A_1 \wedge A_2 \wedge A_3 \to A_3 \wedge A_1 \wedge A_2$ and $\tau_{231} : A_1 \wedge A_2 \wedge A_3 \to A_2 \wedge A_3 \wedge A_1$ are the obvious twisting maps.

Then, we state:

Proposition 3.7. 1. *If $k[E^nX, EX] = 0$ then $k\pi_n(EX) \subseteq \mathrm{G}_n(EX)$ for $n \geq 2$. In particular, if $M(A,m)$ is the Moore space of type (A,m) and $k[M(A,m+n-1), M(A,m)] = 0$ then $k\pi_n(M(A,m)) \subseteq \mathrm{G}_n(M(A,m))$.*

2. *If the suspension map*

$$E : [E^kX, EX] \to [E^{k+1}X, E^2X]$$

is a monomorphism then $G_k(EX) = \pi_k(EX)$ for $k \geq 1$. In particular, if the suspension map

$$E : [M(A,m+k), (M(A,m)] \to [M(A,m+k+1), M(A,m+1)]$$

is a monomorphism then $G_k(M(A,m)) = \pi_k(M(A,m))$ for $k \geq 1$.

3. $4\pi_k(M^n) \subseteq G_k(M^n) \subseteq P_k(M^n)$ *for* $k \geq 1$ *and* $n \geq 3$.
4. *If* $\alpha \in G_k(M^n)$ *then* $\alpha\beta \in G_m(M^n)$ *for any* $\beta \in \pi_m(\mathbb{S}^k)$.
   If $\alpha \in P_k(M^n)$ *then* $\alpha\beta \in P_m(M^n)$ *for any* $\beta \in \pi_m(\mathbb{S}^k)$.
5. *If* $\alpha \in \pi_k(M^n)$ *with* $\sharp\alpha = 2$ *then* $[\alpha, \alpha] \in P_{2k-1}(M^n)$ *for* $k \geq 2$ *and* $n \geq 3$.
   In particular, $[i_n, i_n] \in P_{2n-3}(M^n)$ *for* $n \geq 3$.

Proof. 1. If $\alpha \in \pi_n(EX)$ then $[k\alpha, \iota_{EX}] \in k[E^n X, EX] = 0$ for $n \geq 2$.
2. This is a consequence of (3.4).
3. This follows from (1), (1.41) and (3.2).
4. If $[\iota_{M^n}, \alpha] = 0$ for $\alpha \in \pi_k(M^n)$ then (3.2) implies that $[\iota_{M^n}, \alpha\beta] = 0$ for $\beta \in \pi_m(\mathbb{S}^k)$. Similarly, if $[\gamma, \alpha] = 0$ for $\alpha \in \pi_k(M^n)$ and $\gamma \in \pi_l(M^n)$, then (3.2) leads to $[\gamma, \alpha\beta] = 0$ for $\beta \in \pi_m(\mathbb{S}^k)$.
5. If $\alpha \in \pi_k(M^n)$ and $\beta \in \pi_m(M^n)$ then the Jacobi identity implies

$$[[\alpha, \alpha], \beta] \pm [[\alpha, \beta], \alpha] \pm [[\beta, \alpha], \alpha] = 0.$$

Consequently, $[[\alpha, \alpha], \beta] = 0$ for any $\beta \in \pi_m(M^n)$ provided $\sharp\alpha = 2$ and the proof follows.

□

To state the next results, we use the notions from Sects. 1.2–1.4. In particular, we recall that there is the cell structure $V_{2n+1,2} = M^{2n} \cup_{\lambda_n} e^{4n-1}$ for some $\lambda_n \in \pi_{4n-2}(M^{2n})$ for $n \geq 2$, where the order of λ_n is 4 for n even and 8 for n odd.

In view of [67, Theorem 1.2], if n is not a power of 2 then $[i_n, i_n] \neq 0$.

Next, we recall that in view of [58, Lemma 1.5(ii),(iii)], it holds:

$$\begin{aligned} &1.\ \pi_{2n+1}(E(M^n \wedge M^n)) = \mathbb{Z}_2\{\tilde{\iota}_{2n}\eta_{2n}\} \oplus \mathbb{Z}_2\{i'_{2n}\tilde{\eta}_{2n-1}\},\\ &2.\ \pi_{2n+2}(EM^n \wedge M^n) = \mathbb{Z}_2\{\tilde{\iota}_{2n}\eta^2_{2n}\} \oplus \mathbb{Z}_2\{i'_{2n}\tilde{\eta}_{2n-1}\eta_{2n+1}\} \oplus \mathbb{Z}_2\{i''_{2n}\nu_{2n-1}\} \end{aligned} \tag{3.5}$$

for $n \geq 3$.

Now, we are in a position to show:

Lemma 3.8. *If* $n \geq 2$ *then* $[i_{2n}, \tilde{\eta}_{2n-1}] = \lambda_n\eta_{4n-2} \neq 0$ *and* $\lambda_n\eta^2_{4n-2} = [i_{2n}, \tilde{\eta}_{2n-1}]\eta_{4n-1} \neq 0$.

Proof. We make use of the EHP-sequence

$$\cdots \to \pi_k(M^n) \xrightarrow{E} \pi_{k+1}(M^{n+1}) \xrightarrow{H} \pi_{k+1}(E(M^n \wedge M^n)) \xrightarrow{P} \pi_{k-1}(M^n) \xrightarrow{E} \pi_k(M^{n+1}) \to \cdots.$$

Then, (1.98) and (3.5)(2) lead to $H(\lambda_n\eta^2_{4n-2}) = \tilde{\iota}_{4n-2}\eta^2_{4n-2} \in \pi_{4n}(EM^{2n-1} \wedge M^{2n-1})$. Consequently, we get $\lambda_n\eta^2_{4n-2} \neq 0$ which implies $\lambda_n\eta_{4n-2} \neq 0$.

Next, (1.99) implies $P(\tilde{\iota}_{4n}\eta_{4n}) = 0$ and (1.101)(1) leads to

$$P(i'_{4n}\tilde{\eta}_{4n-1}) = P(i'_{4n})\tilde{\eta}_{4n-3} = [\iota_{M^{2n}}, i_{2n}]\tilde{\eta}_{4n-3} = [i_{2n}, \tilde{\eta}_{2n-1}]$$

Hence, (3.5)(1) yields $P\pi_{4n+1}(E(M^{2n} \wedge M^{2n})) = \{[i_{2n}, \tilde{\eta}_{2n-1}]\}$.

Because $E\lambda_n = i_{2n+1}[\iota_{2n}, \iota_{2n}]$ (1.97), in view of [58, Lemma 2.4(ii)], it holds $E(\lambda_n(\eta_{4n-2})) = i_{2n+1}[\iota_{2n}, \eta_{2n}] = 0$. Then, we deduce that $\lambda_n\eta_{4n-2} \in P\pi_{4n+1}(E(M^{2n} \wedge M^{2n}))$. This implies $[i_{2n}, \tilde{\eta}_{2n-1}] = \lambda_n\eta_{4n-2} \neq 0$ and $[i_{2n}, \tilde{\eta}_{2n-1}]\eta_{4n-1} = \lambda_n\eta^2_{4n-2}$, and the proof is complete. □

Problem 3.9. Does the following hold:
a suitable choice of an extension $\overline{[\iota_{2n-1}, \iota_{2n-1}]}$ of $[\iota_{2n-1}, \iota_{2n-1}]$ implies $[\iota_{M^{2n}}, i_{2n}] = (\lambda_n)p_{4n-2} + i_{2n}\overline{[\iota_{2n-1}, \iota_{2n-1}]}$ for $n \geq 3$?

We show:

Proposition 3.10. 1. *$P_{n-1}(M^n) = 0$ for $n \geq 3$;*
2. *$P_3(M^3) = 2\pi_3(M^3) = \mathbb{Z}_2\{2i_3\eta_2\}$ and $P_n(M^n) = 0$ for with $n \geq 4$;*
3. *$\tilde{\eta}_{n-1} \notin P_{n+1}(M^n)$ for $n \geq 3$ and $P_{n+1}(M^n) = 0$ for n odd and $n \geq 3$;*
4. *$P_{n+2}(M^n) = 0$, $P_{n+3}(M^n) = 0$ for n odd with $n \geq 7$ and $P_{n+4}(M^n) = 0$ for $n \geq 8$.*

Proof. First, recall that $[i_3\eta_2, \tilde{\eta}_2]$ is one of generators of the group $\pi_6(M^3) \cong (\mathbb{Z}_2)^5$ [31, Lemma 3.6] and $[\tilde{\eta}_4, i_5]\eta_9$ is the generator of the direct summand $\mathbb{Z}_2$ of the group $\pi_{10}(M^5) \cong \mathbb{Z}_4 \oplus \mathbb{Z}_2$ [43, Theorem 1.2]. Further, $[i_n, \tilde{\eta}_{n-1}]\eta^2_{2n-1} \neq 0$ [58, Lemma 3.3(i)] for an odd n with $n \geq 7$, and $[i_{2n}, \tilde{\eta}_{2n-1}]\eta_{4n-1} \neq 0$ for $n \geq 2$ (Lemma 3.8).

(1)–(3) The above and (3.3) yield that $i_n\eta_{n-1} \notin P_n(M^n)$ for $n \geq 3$ and $P_n(M^n) = 0$ for $n \geq 4$. Because $[i_3\eta_2, \iota_{M^3}] \in [M^5, M^3]$ and $2[M^5, M^3] = 0$ [31, Lemma 3.7(i)], we get $[2i_3\eta_2, \iota_{M^3}] = 0$. Thus, in view of (3.2) and Proposition 3.7, we get $P_3(M^3) = 2\pi_3(M^3) \cong \mathbb{Z}_2$. Further, (3.2) leads to $[i_n, \tilde{\eta}_{n-1}] \neq 0$ for $n \geq 3$. Consequently, $P_{n-1}(M^n) = 0$ and $\tilde{\eta}_{n-1} \notin P_{n+1}(M^n)$ for $n \geq 3$. In view of [58, Theorem 4.1(i)], the order $\sharp[\tilde{\eta}_{n-1}, \tilde{\eta}_{n-1}] = 4$ if n is odd. Therefore, $2\tilde{\eta}_{n-1} \notin P_{n+1}(M^n)$ and $P_{n+1}(M^n) = 0$ if n is odd.

(4) Lemma 1.27(3) yields that $P_{n+4}(M^n) = 0$ for $n \geq 8$. To show that $P_{n+2}(M^n) = 0$, $P_{n+3}(M^n) = 0$ for n odd and $n \geq 7$, we recall that in view of Lemma 1.27(1), we have $\pi_{n+2}(M^n) = \mathbb{Z}_2\{\tilde{\eta}_{n-1}\eta_{n+1}\} \oplus \mathbb{Z}_2\{i_n\nu_{n-1}\}$ for $n \geq 6$ and $\pi_{n+3}(M^n) = \mathbb{Z}_2\{\tilde{\eta}_{n-1}\eta^2_{n+1}\}$ for $n \geq 7$. Then, $[i_n, \tilde{\eta}_{n-1}]\eta^2_{2n-1} \neq 0$ for an odd n with $n \geq 7$ [58, Lemma 3.3(i)] implies that $\tilde{\eta}_{n-1}\eta_{n+1} \notin P_{n+2}(M^n)$ and $\tilde{\eta}_{n-1}\eta^2_{n+1} \notin P_{n+3}(M^n)$.

Further, $i_n[\iota_{n-1}, \nu_{n-1}] \neq 0$ for an odd n and $n \geq 5$ [58, Lemma 3.4] yields $i_n\nu_{n-1} \notin P_{n+2}(M^n)$ for n is odd and $n \geq 5$. Hence, $P_{n+2}(M^n) = 0$ and $P_{n+3}(M^n) = 0$ for n odd and $n \geq 7$, and the proof is complete. □

Now, we deduce:

Corollary 3.11. 1. *$G_{n-1}(M^n) = 0$ for $n \geq 3$;*
2. *$G_3(M^3) = 2\pi_3(M^3) = \mathbb{Z}_2\{2i_3\eta_2\}$ and $\mathrm{G}_n(M^n) = 0$ for with $n \geq 4$;*
3. *$G_{n+1}(M^n) = 0$ for $n \geq 3$;*
4. *$G_{n+2}(M^n) = 0$, $G_{n+3}(M^n) = 0$ for n odd with $n \geq 7$ and $G_{n+4}(M^n) = 0$ for $n \geq 8$.*

Proof. In the proof of Proposition 3.10 we have shown that $i_3\eta_2 \notin P_3(M^3)$ and $[2i_3\eta_2, \iota_{M^3}] = 0$. Hence, we derive that $G_3(M^3) = P_3(M^3) = 2\pi_3(M^3) = \mathbb{Z}_2\{2i_3\eta_2\}$. Because $G_k(M^n) \subseteq P_k(M^n)$, the properties (1), (2) and (4) follow from those of Proposition 3.10.

(3) In view of Proposition 3.10, we have to show that $G_{2n+1}(M^{2n}) = 0$ for $n \geq 2$. Applying (1.40), we obtain $[\iota_{M^{2n}}, 2\tilde{\eta}_{2n-1}] = [i_{2n}\eta_{2n-1}p_{2n}, \tilde{\eta}_{2n-1}] = [i_{2n}, \tilde{\eta}_{2n-1}]\eta_{4n-1}p_{4n}$. Because $[i_{2n}, \tilde{\eta}_{2n-1}]\eta_{4n-1} \neq 0$ for $n \geq 2$ (Lemma 3.8), the cofiber sequence $(\mathcal{CS}_{4n})$ leads to $[\iota_{M^{2n}}, 2\tilde{\eta}_{2n-1}] \neq 0$ and consequently $G_{2n+1}(M^{2n}) = 0$ for $n \geq 2$. Then, Proposition 3.10(3) completes the proof. □

Remark 3.12. 1. Notice that Proposition 3.10 yields another proof of the result $G_{n-1}(M^n) = 0$ for $n \geq 3$ stated in Proposition 3.3.
2. Because $[2\iota_{M^n}, \tilde{\eta}_{n-1}] \neq 0$ for $n \geq 3$ and M^n is an abelian H'-space for $n \geq 4$, the relations (3.2) and (1.41) imply that the order $\sharp[\iota_{M^n}, \iota_{M^n}] = 4$ for $n \geq 4$.

3.2 Gottlieb Groups of Some Moore Spaces $M(A, n)$

If A is an abelian group and $M(A, n)$ is the Moore space of type (A, n) for $n \geq 2$ then Proposition 3.3 yields that $G_n(M(A, n)) = 0$, if n is even or n is odd and A is torsion. Now, we examine $G_{n+k}(M(A, n)$ for $k \leq 4$.

Let $j_1 : X \hookrightarrow X \vee Y$ and $j_2 : Y \hookrightarrow X \vee Y$ be the inclusion maps. Then, we recall:

Lemma 3.13 ([5, Lemma 4.1]). *Let X and Y be spaces such that $\pi_k(X \vee Y) = j_{1*}\pi_k(X) \oplus j_{2*}\pi_k(Y)$ and $G_k(X) = 0$. Then $G_k(X \vee Y) \subseteq j_{2*}G_k(Y)$.*

According to [5, Remark 4.2], we can state:

Remark 3.14. If spaces X, Y are $(m-1)$- and $(n-1)$-connected, respectively, then $\pi_{k+1}(X \times Y, X \vee Y) = 0$ for $k + 1 < m + n$. Hence, the hypothesis above on $\pi_k(X \vee Y)$ holds provided $k + 1 < m + n$. In particular, for $X = M(A, m)$ and $Y = M(B, n)$.

To state the next result, we need:

Lemma 3.15. *If A, B are torsion abelian groups whose primary components are determined by disjoint sets of primes then the inclusion map $M(A, m) \vee M(B, n) \hookrightarrow M(A, m) \times M(B, n)$ is a homotopy equivalence as a homology isomorphism for $m, n \geq 2$.*

Proof. The Künneth formula for homology groups of the pairs of pointed spaces (X, x_0) and $Y, y_0)$ gives

$$H_k(X \times Y, X \vee Y) = \bigoplus_{i=0}^{n} H_i(X, x_0) \otimes H_{n-i}(Y, y_0) \oplus \bigoplus_{i=0}^{k} \mathrm{Tor}(H_i(X, x_0), H_{k-i-1}(Y, y_0)).$$

If primary components of groups A and B are determined by disjoint sets of primes then

$$H_i(M(A,m),x_0) \otimes H_{k-i}(M(B,n),y_0) = 0$$

and

$$\mathrm{Tor}(H_i(M(A,m),x_0), H_{k-i-1}(M(B,n),y_0)) = 0$$

for fixed points $x_0 \in M(A,m)$ and $y_0 \in M(B,n)$. Hence, the exact homology sequence of the pair $(M(A,m) \times M(B,n), M(A,m) \vee M(B,n))$ implies that the inclusion map $M(A,m) \vee M(B,n) \hookrightarrow M(A,m) \times M(B,n)$ is a homology isomorphism. Consequently, $M(A,m) \vee M(B,n) \hookrightarrow M(A,m) \times M(B,n)$ is also a homotopy equivalence and the proof is complete. □

Remark 3.16. Notice that Lemma 3.15 implies that the space $M(A,m) \wedge M(B,n)$ is contractible for $m,n \geq 2$ provided primary components of A and B are determined by disjoint sets of primes. In particular, the space $M(\mathbb{Z}_k,m) \wedge M(\mathbb{Z}_l,n)$ is contractible for $m,n \geq 2$ if k and l are relatively prime.

Then, we can show:

Theorem 3.17. *Let A be an abelian group. Then:*

1. $G_{n-1}(M^m \vee M(A,n)) \subseteq j_{1*}G_{n-1}(M^m)$ *for* $m \geq 3$ *and* $n \geq 2$;
2. $G_{m-1}(M^m \vee M(A,n)) \subseteq j_{2*}G_{m-1}(M(A,n))$ *for* $m \geq 3$ *and* $n \geq 2$;
3. $G_m(M^m \vee M(A,n)) \subseteq j_{2*}G_m(M(A,n))$ *for* $m \geq 4$ *and* $n \geq 3$;
4. $G_{m+1}(M^m \vee M(A,n)) \subseteq j_{2*}G_{m+1}(M(A,n))$ *for* $m \geq 3$ *and* $n \geq 4$;
5. $G_{m+2}(M^m \vee M(A,n)) \subseteq j_{2*}G_{m+2}(M(A,n))$ *for m odd with* $m \geq 7$ *and* $n \geq 5$;
6. $G_{m+3}(M^m \vee M(A,n)) \subseteq j_{2*}G_{m+3}(M(A,n))$ *for m odd with* $m \geq 7$ *and* $n \geq 6$;
7. $G_{m+4}(M^m \vee M(A,n)) \subseteq j_{2*}G_{m+4}(M(A,n))$ *for* $m,n \geq 8$;
8. $G_k(M(A,m) \vee M(B,n)) = j_{1*}G_k(M(A,m)) \oplus j_{2*}G_k(M(B,n)$ *for* $k \geq 1$ *and* $m,n \geq 2$ *if* A, B *are torsion abelian groups whose primary components are determined by disjoint sets of primes. In particular,* $G_k(M^m \vee M(A,n)) = j_{1*}G_k(M^m) \oplus j_{2*}G_k(M(A,n))$ *for all* $k \geq 1$, $m \geq 3$ *and* $n \geq 2$ *if A is a torsion group whose two-primary component is trivial.*

Proof. (1) Because M^m with $m \geq 3$ is $(m-2)$-connected and $M(A,n)$ with $n \geq 2$ is $(n-1)$-connected, by Remark 3.14, we get $\pi_k((M^m \vee M(A,n)) = j_{1*}\pi_k(M^m) \oplus j_{2*}\pi_k(M(A,n))$ provided $k+2 < m+n$. But $\pi_{n-1}(M(A,n)) = G_{n-1}(M(A,n)) = 0$,so Lemma 3.13 leads to $G_{n-1}(M^m \vee M(A,n)) \subseteq j_{1*}G_{n-1}(M^m)$.

(2)–(7) Those are direct consequences of Remark 3.14, Corollary 3.11 and Lemma 3.13.

Nevertheless, we present the proof of (7) by Remark 3.14 we have $\pi_{m+4}((M^m \vee M(A,n)) = \pi_{m+4}(M^m) \oplus \pi_{m+4}(M(A,n))$ provided $n \geq 8$. Then, Corollary 3.11 and Lemma 3.13 lead to $\mathrm{G}_{m+4}(M^m \vee M(A,n)) \subseteq j_{2*}\mathrm{G}_{m+4}(M(A,n))$ for $m,n \geq 8$.

(8) This follows from Lemma 3.15 and the Gottlieb's result [22, Theorems 1-7, 2-1]. □

Further, we derive:

Corollary 3.18. 1. $G_{n-1}(M^m \vee M(A,n)) = 0$ *for* $m \geq n+1$ *and* $n \geq 2$;
2. $G_{n-1}(M^n \vee M^{n+k}) = 0$ *for* $k \geq 0$ *and* $n \geq 3$;
3. $G_n(M^n \vee M^{n+k}) = 0$ *for* $k \geq 0$ *and* $n \geq 4$;
4. $G_{n+1}(M^n \vee M^{n+k}) = 0$ *for* $k \geq 0$ *and* $n \geq 3$;
5. $G_{n+2}(M^n \vee M^{n+k}) = 0$ *and* $G_{n+3}(M^m \vee M^{n+k}) = 0$ *for* $k \geq 0$ *and* n *odd with* $m \geq 7$;
6. $G_{n+4}(M^n \vee M^{n+k}) = 0$ *for* $k \geq 0$ *and* $n \geq 8$.

Notice that Corollary 3.11 and Corollary 3.18(2)–(6) imply that $G_{n+k}(M(\mathbb{Z}_2^t, n)) = G_{n+k}(\underbrace{M^{n+1} \vee \cdots \vee M^{n+1}}_{t}) = 0$ for $k = 1,2,3,4,5$ and any $t \geq 0$ (under some restrictions on n), where $\mathbb{Z}_2^t$ is the direct sum of t-copies of the cyclic group $\mathbb{Z}_2$.

Proposition 3.19. *Let A be a finite abelian group with odd order . Then:*

1. $\pi_{n+1}(M(A,n)) = 0$ *for* $n \geq 3$;
2. $\pi_{n+2}(M(A,n)) = 0$ *for* $n \geq 4$;
3. $\pi_{n+3}(M(A,n)) = 0$ *for* $n \geq 5$ *and* $\pi_{n+4}(M(A,n)) = 0$ *for* $n \geq 6$ *provided the three-primary component of A is trivial;*
4. $\pi_{n+5}(M(A,n)) = 0$ *for* $n \geq 7$;
5. $G_{n+k}(M(A \oplus \mathbb{Z}, n)) \subseteq j_{2*}G_{n+k}(\mathbb{S}^n)$ *provided* $n \geq k+4$ *for* $k = 0,1,2,3,4,5$ *and the restrictions from* (1)–(6) *above.*

Proof. Let $A = A_{p_1} \oplus \cdots \oplus A_{p_l}$, where A_{p_i} is the p_i-primary component of A for $i = 1,\ldots,l$. Because $M(A,n) = M(A_{p_1},n) \vee \cdots \vee M(A_{p_l},n)$, Theorem 3.17(8) implies that

$$\pi_k(M(A,n)) = j_{1*}\pi_k(M(A_{p_1},n) \oplus \cdots \oplus j_{l*}\pi_k(M(A_{p_l},n))$$

for $k \geq 1$. Therefore, we may assume that A is a p-primary finite abelian group with $p \geq 3$ and let

$$A = \mathbb{Z}_{p^{s_1}} \oplus \cdots \oplus \mathbb{Z}_{p^{s_t}}$$

with $s_1,\ldots,s_t > 0$. Since $M(A,n) = M(\mathbb{Z}_{p^{s_1}},n) \vee \cdots \vee M(\mathbb{Z}_{p^{s_t}},n)$ is the $(2n-1)$-skeleton of $M(\mathbb{Z}_{p^{s_1}},n) \times \cdots \times M(\mathbb{Z}_{p^{s_t}},n)$, [4, Proposition 1.5.24] yields an isomorphism

$$\begin{aligned}\pi_k(M(A,n)) &= \pi_k(M(\mathbb{Z}_{p^{s_1}},n) \vee \cdots \vee M(\mathbb{Z}_{p^{s_t}},n)) \\ &\xrightarrow{\cong} j_{1*}\pi_k(M(\mathbb{Z}_{p^{s_1}},n)) \oplus \cdots \oplus j_{t*}\pi_k(M(\mathbb{Z}_{p^{s_t}},n))\end{aligned}$$

for $k \leq 2n-2$.

(1) and (2) In view of [4, Proposition 4.6.11], we get $\pi_{n+1}(M(A_{p_i}, n)) = 0$ for $n \geq 3$ and $\pi_{n+2}(M(A_{p_i}, n)) = 0$ for $n \geq 4$ and $i = 1, \ldots, t$.

(3) First, if $\mathcal{Z}_{\mathbf{m}}$ is the mapping cylinder of a map $\mathbf{m} : \mathbb{S}^n \to \mathbb{S}^n$ of degree m then we have a cofiber sequence

$$\mathbb{S}^n \longrightarrow \mathcal{Z}_{\mathbf{m}} \longrightarrow M(\mathbb{Z}_m, n).$$

Then, by Blakers–Massey theorem [4, Theorem 5.6.4], we obtain the exact sequence

$$\pi_k(\mathbb{S}^n) \overset{\mathbf{m}_*}{\to} \pi_k(\mathbb{S}^n) \overset{j_*}{\to} \pi_k(M(\mathbb{Z}_m, n)) \overset{d_k}{\to} \pi_{k-1}(\mathbb{S}^n) \overset{\mathbf{m}_*}{\to} \pi_{k-1}(\mathbb{S}^n),$$

where $j : \mathbb{S}^n \hookrightarrow M(\mathbb{Z}_m, n)$ is the inclusion map, d_k is the boundary homomorphism, and $k \leq 2n-2$. Because $\pi_{n+2}(\mathbb{S}^n) = \mathbb{Z}_2\{\eta_n^2\}$ for $n \geq 2$, $\pi_{n+3}(\mathbb{S}^n) = \mathbb{Z}_{24}\{\nu_n^+\}$ for $n \geq 5$ and $\pi_{n+4}(\mathbb{S}^n) = 0$ for $n \geq 6$, Blakers–Massey theorem and the comments above lead to $\pi_{n+3}(M(A_{p_i}, n)) = 0$ for $n \geq 5$ and $\pi_{n+4}(M(A_{p_i}, n)) = 0$ for $n \geq 6$ and $i = 1, \ldots, t$ provided that the three-primary component of A is trivial.

(4) Since $\pi_{n+4}(\mathbb{S}^n) = 0$ for $n \geq 6$ and $\pi_{n+5}(\mathbb{S}^n) = 0$ for $n \geq 7$, again Blakers–Massey theorem shows that $\pi_{n+5}(M(A_{p_i}, n)) = 0$ for $n \geq 7$ and $i = 1, \ldots, t$.

(5) This follows from Lemma 3.15 and Theorem 3.17 and the proof is complete. □

Notice that Theorem 3.17(8) implies that

$$G_k(M(A, n)) = j_{1*}G_k(M(A_{p_1}, n)) \oplus \cdots \oplus j_{l*}G_k(M(A_{p_l}, n))$$

for $k \geq 1$ if $A = A_{p_1} \oplus \cdots \oplus A_{p_l}$, where A_{p_i} is the p_i-primary component of A for $i = 1, \ldots, l$. In particular, if A is a finite abelian group. Further, in view of [5, Corollary 4.4] it holds $lj_{1*}G_n(\mathbb{S}^n) \subseteq G_n(M(A \oplus \mathbb{Z}, n)) \subseteq j_{1*}G_n(\mathbb{S}^n)$ for n odd, a finite abelian group A and some even l. But, Proposition 3.19 leads to:

Corollary 3.20. 1. $G_{n+k}(M(A \oplus \mathbb{Z}, n)) = j_{2*}G_{n+k}(\mathbb{S}^n)$ *provided* $n \geq k+4$ *for* $k = 1, 2$ *and the order of* A *is odd;*

2. $G_{n+3}(M(A \oplus \mathbb{Z}), n) = j_{2*}G_{n+3}(\mathbb{S}^n)$ *for* n *odd with* $n \geq 7$ *provided the order of*
A is odd and the three-primary component of A is trivial;
3. $G_{n+4}(M(A \oplus \mathbb{Z}), n) = j_{2*}G_{n+4}(\mathbb{S}^n)$ *for* n *odd with* $n \geq 9$ *provided the order of*
A is odd and the three-primary component of A is trivial;
4. $G_{n+5}(M(A \oplus \mathbb{Z}), n) = j_{2*}G_{n+5}(\mathbb{S}^n)$ *for* $n \geq 8$ *provided the order of* A *is odd.*

Proof. In view of Proposition 3.19, the results (2)–(5) hold provided $G_{n+k}(\mathbb{S}^n) = 0$ for $k = 1, 2, 3, 4, 5$.

To consider the case $G_{n+k}(\mathbb{S}^n) \neq 0$ for $k = 1, 2, 3, 4, 5$, we recall (see, e.g., [86, Chap. V]) that $\pi_{n+1}(\mathbb{S}^n) = \mathbb{Z}_2\{\eta_n\}$ for $n \geq 3$, $\pi_{n+2}(\mathbb{S}^n) = \mathbb{Z}_2\{\eta_n^2\}$ for $n \geq 2$, $\pi_{n+3}(\mathbb{S}^n) = \mathbb{Z}_{24}\{\nu_n^+\}$ for $n \geq 5$, $\pi_{n+4}(\mathbb{S}^n) = 0$ for $n \geq 6$ and $\pi_{n+5}(\mathbb{S}^n) = 0$ for $n \geq 7$. Further, by means of the universal coefficient theorem for homotopy [27, p. 30], for any pointed space X, an abelian group A and $n \geq 2$, there exists a short exact sequence

$$0 \longrightarrow \text{Ext}(A, \pi_{n+1}(X)) \longrightarrow [M(A,n), X] \longrightarrow \text{Hom}(A, \pi_n(X)) \longrightarrow 0.$$

Now, if the order of A is odd and $X = M(A \oplus \mathbb{Z}, n)$ then the short exact above implies that $j_2 : M(A,n) \hookrightarrow M(A \oplus \mathbb{Z}, n)$ has an odd order (respectively non-divisible by 3) in the group $[M(A,n), M(A \oplus \mathbb{Z}, n)]$ provided the order of A is odd (respectively the three-primary component of A is trivial as well). Consequently, $[j_1\eta_n, j_2] = [j_1\eta_n^2, j_2] = [j_1\nu_n^+, j_2] = 0$. Because $[j_1\eta_n, j_1] = j_1[\eta_n, \iota_n] = 0$, $[j_1\eta_n^2, j_1] = j_1[\eta_n^2, \iota_n] = 0$ and $[j_1 l\nu_n^+, j_1] = j_1[l\nu_n^+, \iota_n] = 0$ if $\eta_n \in G_{n+1}(\mathbb{S}^n)$, $\eta_n^2 \in G_{n+2}(\mathbb{S}^n)$ and $l\nu_n^+ \in G_{n+3}(\mathbb{S}^n)$ for some $1 \leq l \leq 23$ (see Chap. 1 for the details) the proof is complete. □

We close the volume with the two types of tables. First, the table of the order of $[\iota_n, \alpha]$, where $\alpha \in \pi_{n+k}^n$ for $n \geq k+2$, $k \leq 11$ and $n \equiv r \pmod 8$ with $0 \leq r \leq 7$, given except as otherwise noted. This corrects the table in [66, the second page], where $m \equiv n$ (k) indicates $m \equiv n \pmod k$ and symbols in italic stress irregular cases.

Next tables of $G_{n+k}(\mathbb{S}^n)$ for $1 \leq k \leq 13$ and $2 \leq n \leq 26$, $P_{n+k}(\mathbb{R}P^n)$ for $k \leq 13$ and $2 \leq n \leq 26$, and the table of $P_{k+2n+1}(\mathbb{C}P^n)$ for $k \leq 13$ and $2 \leq n \leq 26$ are given by compiling our results. Like in [85, Chap. XIV], an integer n indicates the cyclic group $\mathbb{Z}_n$ of order n, the symbol ∞ the infinite cyclic group $\mathbb{Z}$, the symbol $+$ the direct sum of groups, and $(2)^k$ the direct sum of k-copies of $\mathbb{Z}_2$.

The table of the order of $[\iota_n, \alpha]$

$\alpha \backslash r$	0	1	2	3	4	5	6	7
η	2	2	2	1	2	2	$2, \neq 6$ $1, = 6$	1
η^2	2	2	1	1	2	$2, \neq 5$ 1,=5	1	1
ν	8	2	4	2	$8, \neq 12$ $4, = 12$	$2, \neq 2^i - 3$ $1, = 2^i - 3$	4	1
ν^2	2	2	2	$2, \neq 2^i - 5$ $1, = 2^i - 5$	1	1	2	1
σ	16	2	16	$2, \neq 11$ $1, = 11$	16	2	16	$2, 7(16)$ $1, 15(16)$
$\eta\sigma$	2	2	$2, \neq 10$ $1, = 10$	1	2	2	$2, \equiv 22(32)$ ≥ 54 1, otherwise	1
ε	2	2	1	1	2	2	2	1
$\bar{\nu}$	2	2	$2, \neq 10$ $1, = 10$	1	2	2	2	1
$\eta^2\sigma$	2	$2, \neq 2^i - 7$ $1, = 2^i - 7$	1	1	2	$2, \equiv 53(64)$ ≥ 117 $1, \not\equiv 53(64)$	1	1
$\eta\varepsilon$	2	1	1	1	2	$2, \equiv 53(64)$ ≥ 117 $1, \not\equiv 53(64)$	1	1

(continued)

(continued)

$\alpha \backslash r$	0	1	2	3	4	5	6	7
ν^3	2	$2, \neq 2^i - 7$ $1, = 2^i - 7$	1	1	1	1	1	1
μ	2	2	2	1	2	2	2	1
$\eta\mu$	2	2	1	1	2	2	1	1
ζ	8	1	4	$2, \equiv 115(128)$ ≥ 243 $1, \not\equiv 115(128)$	8	1	4	1

The table of $G_{n+k}(\mathbb{S}^n)$ for $1 \leq k \leq 13$ and $2 \leq n \leq 26$

$G_{n+k}(\mathbb{S}^n)$	$n = 2$	$n = 3$	$n = 4$	$n = 5$	$n = 6$	$n = 7$	$n = 8$
$k = 1$	∞	2	0	0	2	2	0
$k = 2$	2	2	0	2	2	2	0
$k = 3$	2	12	$3\infty + 2$	24	2	24	0
$k = 4$	12	2	$(2)^2$	2	0	0	0
$k = 5$	2	2	$(2)^2$	2	3∞	0	0
$k = 6$	2	3	$24 + 3$	2	0	2	0
$k = 7$	3	15	0	30	0	120	$3\infty + 2$
$k = 8$	15	2	0	0	$24 + 2$	$(2)^3$	$(2)^2$
$k = 9$	2	$(2)^2$	2	$(2)^2$	$(2)^3$	$(2)^4$	$(2)^2$
$k = 10$	$(2)^2$	$12 + 2$	$120 + 6$	$72 + 2$	$24 + 2$	$24 + 2$	$24 + 8$
$k = 11$	$12 + 2$	$84 + (2)^2$	$(2)^6$	$504 + (2)^2$	$4 + 2$	$504 + 2$	2
$k = 12$	$84 + (2)^2$	$(2)^2$	$(2)^6$	$(2)^3$	240	0	0
$k = 13$	$(2)^2$	6	$24 + (2)^2$	$6 + 2$	2	6	$(2)^2$

$G_{n+k}(\mathbb{S}^n)$	$n=9$	$n=10$	$n=11$	$n=12$	$n=13$	$n=14$	$n=15$	$n=16$	$n=17$
$k=1$	0	0	2	0	0	0	2	0	0
$k=2$	0	2	2	0	0	2	2	0	0
$k=3$	12	2	12	2	24	2	24	0	12
$k=4$	0	0	0	0	0	0	0	0	0
$k=5$	0	0	0	0	0	0	0	0	0
$k=6$	0	0	2	2	2	0	2	0	0
$k=7$	120	0	240	0	120	0	240	0	120
$k=8$	2	$(2)^2$	$(2)^2$	0	0	2	$(2)^2$	0	0
$k=9$	$(2)^3$	$3\infty+(2)^2$	$(2)^3$	2	$(2)^2$	$(2)^2$	$(2)^3$	0	2
$k=10$	24	$4+2$	$6+2$	0	3	2	6	0	3
$k=11$	$504+2$	2	504	3∞	504	2	504	0	504
$k=12$	0	4	2	$(2)^2$	2	0	0	0	0
$k=13$	6	2	$6+2$	$(2)^2$	6	3∞	3	0	3

$G_{n+k}(\mathbb{S}^n)$	$n = 18$	$n = 19$	$n = 20$	$n = 21$	$n = 22$	$n = 23$	$n = 24$	$n = 25$	$n = 26$
$k = 1$	0	2	0	0	0	2	0	0	0
$k = 2$	2	2	0	0	2	2	0	0	2
$k = 3$	2	12	0	12	2	24	0	12	2
$k = 4$	0	0	0	0	0	0	0	0	0
$k = 5$	0	0	0	0	0	0	0	0	0
$k = 6$	0	0	2	2	0	2	0	0	0
$k = 7$	0	120	0	120	0	120	0	120	0
$k = 8$	2	$(2)^2$	0	0	2	$(2)^2$	0	0	2
$k = 9$	$(2)^2$	$(2)^3$	2	$(2)^2$	$(2)^2$	$(2)^3$	0	2	$(2)^2$
$k = 10$	2	6	0	3	2	6	0	3	2
$k = 11$	2	504	0	504	2	504	0	504	2
$k = 12$	0	0	0	0	0	0	0	0	0
$k = 13$	0	3	0	3	0	3	0	3	0

The table of $P_{n+k}(\mathbb{R}P^n)$ for $k \leq 13$ and $2 \leq n \leq 26$

$P_{n+k}(\mathbb{R}P^n)$	$n = 2$	$n = 3$	$n = 4$	$n = 5$	$n = 6$	$n = 7$	$n = 8$
$k = 0$	0	∞	0	2∞	0	∞	0
$k = 1$	∞	2	0	0	2	2	0
$k = 2$	2	2	0	2	2	2	0
$k = 3$	2	12	$3\infty + 2$	24	12	24	0
$k = 4$	12	2	2	2	0	0	0
$k = 5$	2	2	2	2	3∞	0	0
$k = 6$	2	3	8	2	0	2	0
$k = 7$	3	15	0	30	0	120	$3\infty + 2$
$k = 8$	15	2	0	0	$24 + 2$	$(2)^3$	2
$k = 9$	2	$(2)^3$	2	$(2)^2$	$(2)^3$	$(2)^4$	2
$k = 10$	$(2)^2$	$12 + 2$	$120 + 2$	$72 + 2$	$(2)^2$	$24 + 2$	24
$k = 11$	$12 + 2$	$84 + (2)^2$	$(2)^4$	$504 + (2)^2$	$4 + 2$	$504 + 2$	2
$k = 12$	$84 + (2)^2$	$(2)^2$	$(2)^4$	$(2)^3$	240	0	0
$k = 13$	$(2)^2$	6	$8 + 2$	$6 + 2$	2	6	2

$P_{n+k}(\mathbb{R}P^n)$	$n=9$	$n=10$	$n=11$	$n=12$	$n=13$	$n=14$	$n=15$	$n=16$	$n=17$
$k=0$	2∞	0	2∞	0	2∞	0	2∞	0	2∞
$k=1$	0	0	2	0	0	0	2	0	0
$k=2$	0	2	2	0	0	2	2	0	0
$k=3$	2	12	2	12	24	12	24	0	2
$k=4$	0	0	0	0	0	0	0	0	0
$k=5$	0	0	0	0	0	0	0	0	0
$k=6$	0	0	2	2	2	0	2	0	0
$k=7$	2	0	240	0	2	0	240	0	2
$k=8$	2	$(2)^2$	$(2)^2$	0	0	2	$(2)^2$	0	0
$k=9$	$(2)^3$	$3\infty+(2)^2$	$(2)^3$	2	$(2)^2$	$(2)^2$	$(2)^3$	0	2
$k=10$	24	$(2)^2$	$6+2$	0	3	2	6	0	3
$k=11$	504	2	504	3∞	504	2	504	0	504
$k=12$	0	4	2	2	2	0	0	0	0
$k=13$	6	2	$6+2$	2	6	3∞	3	0	3

$P_{n+k}(\mathbb{R}P^n)$	$n = 18$	$n = 19$	$n = 20$	$n = 21$	$n = 22$	$n = 23$	$n = 24$	$n = 25$	$n = 26$
$k = 0$	0	2∞	0	2∞	0	2∞	0	2∞	0
$k = 1$	0	2	0	0	0	2	0	0	0
$k = 2$	2	2	0	0	2	2	0	0	2
$k = 3$	12	2	0	2	12	24	0	2	12
$k = 4$	0	0	0	0	0	0	0	0	0
$k = 5$	0	0	0	0	0	0	0	0	0
$k = 6$	0	0	2	2	0	2	0	0	0
$k = 7$	0	2	0	2	0	2	0	2	0
$k = 8$	2	$(2)^2$	0	0	2	$(2)^2$	0	0	2
$k = 9$	$(2)^2$	$(2)^3$	2	$(2)^2$	$(2)^2$	$(2)^3$	0	$(2)^2$	$(2)^2$
$k = 10$	2	6	0	3	2	6	0	3	2
$k = 11$	2	504	0	504	2	504	0	504	2
$k = 12$	0	0	0	0	0	0	0	0	0
$k = 13$	0	3	0	3	0	3	0	3	0

The table of $P_{k+2n+1}(\mathbb{C}P^n)$ for $k \leq 13$ and $2 \leq n \leq 26$

$P_{k+2n+1}(\mathbb{C}P^n)$	$n = 2$	$n = 3$	$n = 4$	$n = 5$	$n = 6$	$n = 7$	$n = 8$
$k = 0$	2∞	∞	2∞	2∞	2∞	2∞	2∞
$k = 1$	0	2	0	2	0	2	0
$k = 2$	0	2	0	2	0	2	0
$k = 3$	24	24	12	12	24	24	12
$k = 4$	2	0	0	0	0	0	0
$k = 5$	2	0	0	0	0	0	0
$k = 6$	2	2	0	2	2	2	0
$k = 7$	30	120	120	240	120	240	120
$k = 8$	2	$(2)^3$	2	$(2)^2$	0	$(2)^2$	0
$k = 9$	2	$(2)^4$	$(2)^3$	$(2)^3$	$(2)^2$	$(2)^3$	2
$k = 10$	24	$24 + 2$	3	$6 + 2$	3	6	3
$k = 11$	$504 + (2)^2$	$504 + 2$	$504 + 2$	504	504	504	504
$k = 12$	$(2)^3$	0	0	2	2	0	0
$k = 13$	$6 + 2$	6	3	$6 + 2$	6	3	3

$P_{k+2n+1}(\mathbb{C}P^n)$	$n=9$	$n=10$	$n=11$	$n=12$	$n=13$	$n=14$	$n=15$	$n=16$	$n=17$
$k=0$	2∞	2∞	2∞	2∞	2∞	2∞	2∞	2∞	2∞
$k=1$	2	0	2	0	2	0	2	0	2
$k=2$	2	0	2	0	0	2	2	0	0
$k=3$	12	12	24	12	12	24	24	12	12
$k=4$	0	0	0	0	0	0	0	0	0
$k=5$	0	0	0	0	0	0	0	0	0
$k=6$	0	2	2	0	2	2	2	0	0
$k=7$	120	120	120	120	120	120	240	120	120
$k=8$	$(2)^2$	0	$(2)^2$	0	$(2)^2$	0	$(2)^2$	0	$(2)^2$
$k=9$	$(2)^3$	$(2)^2$	$(2)^3$	$(2)^2$	$(2)^3$	$(2)^2$	$(2)^3$	2	$(2)^3$
$k=10$	6	3	6	3	6	3	6	3	6
$k=11$	504	504	504	504	504	504	504	504	504
$k=12$	0	0	0	0	0	0	0	0	0
$k=13$	3	3	3	3	3	3	3	3	3

$P_{k+2n+1}(\mathbb{C}P^n)$	$n = 18$	$n = 19$	$n = 20$	$n = 21$	$n = 22$	$n = 23$	$n = 24$	$n = 25$	$n = 26$
$k = 0$	2∞	2∞	2∞	2∞	2∞	2∞	2∞	2∞	2∞
$k = 1$	0	2	0	2	0	2	0	2	0
$k = 2$	0	2	0	2	0	2	0	2	0
$k = 3$	12	24	12	12	12	24	12	12	12
$k = 4$	0	0	0	0	0	0	0	0	0
$k = 5$	0	0	0	0	0	0	0	0	0
$k = 6$	2	2	0	0	2	2	0	0	2
$k = 7$	120	120	120	120	120	240	120	120	120
$k = 8$	0	$(2)^2$	0	$(2)^2$	0	$(2)^2$	0	$(2)^2$	0
$k = 9$	$(2)^2$	$(2)^3$	2	$(2)^3$	$(2)^2$	$(2)^3$	2	$(2)^3$	2
$k = 10$	3	6	3	6	3	6	3	6	3
$k = 11$	504	504	504	504	504	504	504	504	504
$k = 12$	0	0	0	0	0	0	0	0	0
$k = 13$	3	3	3	3	3	3	3	3	3

References

1. Adams, J. F. (1962). Vector fields on spheres. *Annals of Mathematics, 75*, 603–632.
2. Adams, J. F. (1966). On the groups $J(X)$ IV. *Topology, 5*, 21–71; Correction (1968). *Topology, 7*, 331.
3. Arkowitz, M. (1962). The generalized Whitehead product. *Pacific Journal of Mathematics, 12*, 7–23.
4. Arkowitz, M. (2011). *Introduction to homotopy theory*. New York: Springer.
5. Arkowitz, M., & Maruyama, K. (2014). The Gottieb group of a wedge of suspensions. *Journal of the Mathematical Society of Japan, 66*(3), 735–743.
6. Barcus, W. D., & Barratt, M. G. (1958). On the homotopy classification of the extensions of a fixed map. *Transactions of the American Mathematical Society, 88*, 57–74.
7. Barratt, M. G. (1961). Note on a formula due to Toda. *Journal of the London Mathematical Society, 36*, 95–96.
8. Barratt, M. G., James, M. & Stein, N. (1960). Whitehead products and projective spaces. *Journal of Mathematics and Mechanics, 9*, 813–819.
9. Barratt, M. G., Jones, J. D. S., & Mahowald, M. (1987). The Kervaire invariant and Hopf invariant. *Lecture Notes in Mathematics, 1286*, 135–173.
10. Barratt, M. G., & Mahowald, M. E. (1964). The metastable homotopy of $O(n)$. *Bulletin of the American Mathematical Society, 70*, 758–760.
11. Baues, H. J. (1981). *Commutator calculus and groups of homotopy classes*. London Mathematical Society Lecture Note Series (Vol. 50). Cambridge: Cambridge University Press.
12. Blakers, A. L., & Massey, W. S. (1949). The homotopy groups of a triad. *Proceedings of the National Academy of Sciences USA, 35*, 322–328.
13. Borel, A., & Hirzebruch, F. (1959). Characteristic classes and homogeneous spaces II. *American Journal of Mathematics, 81*, 315–382.
14. Bott, R. (1959). The stable homotopy of the classical groups. *Annals of Mathematics, 70*, 313–337.
15. Cohen, R. F., & Neisendorfer, J. A. (1986). Note on the desupending the Adams map. *Mathematical Proceedings of the Cambridge Philosophical Society, 99*, 59–64.
16. Davis, D., & Mahowald, M. (1989). The $SO(n)$-of-origin. *Forum Mathematicum, 1*, 239–250.
17. Feder, S., Gitler, S., & Lam, K. Y. (1977). Composition properties of projective homotopy classes. *Pacific Journal of Mathematics, 68*(1) , 47–61.
18. Gilmore, M. E. (1967). Complex Stiefel manifolds, some homotopy groups and vector fields. *Bulletin of the American Mathematical Society, 73*, 630–633.
19. Golasiński, M., & Gonçalves, D. L. (2001). Postnikov towers and Gottlieb groups of orbit spaces. *Pacific Journal of Mathematics, 197*(2), 291–300.

© Springer International Publishing Switzerland 2014

M. Golasiński, J. Mukai, *Gottlieb and Whitehead Center Groups of Spheres, Projective and Moore Spaces*, DOI 10.1007/978-3-319-11517-7

20. Golasiński, M., & Mukai, J. (2008). Gottlieb groups of spheres. *Topology, 47*, 399–430.
21. Gottlieb, D. (1965). A certain subgroup of the fundamental group. *American Journal of Mathematics, 87*, 840–856.
22. Gottlieb, D. (1969). Evaluation subgroups of homotopy groups. *American Journal of Mathematics, 91*, 729–756.
23. Gottlieb, D. (1975). Witnesses, transgressions, and the evaluation map. *Indiana University Mathematics Journal, 24-9*, 825–836.
24. Hilton, P. J. (1953). Note on the Jacobi identity for Whitehead products. *Annals of Mathematics, 58*, 429–442.
25. Hilton, P. J., (1954). A certain triple Whitehead product. *Proceedings of the Cambridge Philosophical Society, 50*, 189–197.
26. Hilton, P. J., (1955). A note on the P-homomorphism in homotopy groups of spheres. *Proceedings of the Cambridge Philosophical Society, 59*, 230–233.
27. Hilton, P. J., (1965). *Homotopy theory and duality*. New York: Gordon and Breach Science.
28. Hilton, P. J., & Whitehead, J. H. C. (1953). Note on the Whitehead product. *Annals of Mathematics, 58*, 429–442.
29. Hirato, Y., Kachi, H., & Mimura, M. (2001). Homotopy groups of the homogenous spaces F_4/G_2 and $F_4/Spin(9)$. *JP Journal of Geometry and Topology, 1*(1) , 59–109.
30. Hoo, C. S., & Mahowald, M. (1965). Some homotopy groups of Stiefel manifolds. *Bulletin of the American Mathematical Society, 71*, 661–667.
31. Inoue, T., Miyauchi, T., & Mukai, J. (2012). Self-homotopy of a suspension of the real 5-projective space. *JP Journal of Geometry and Topology, 12–2*, 111–158.
32. James, I. M. (1957). On the suspension sequence. *Annals of Mathematics, 65*, 74–107.
33. James, I. M. (1959). Spaces associated with Stiefel manifolds. *Proceedings of the London Mathematical Society (3), 9*, 115–140.
34. James, I. M. (1960). On H-spaces and their homotopy groups. *Quarterly Journal of Mathematics Oxford Series (2), 11*, 161–179.
35. Kachi, H. (1968). On the homotopy groups of rotation groups R_n. *Journal of the Faculty of Science, Shinshu University, 3*, 13–33.
36. Kachi, H., & Mukai, J. (1999). Some homotopy groups of rotation groups R_n. *Hiroshima Mathematical Journal, 29*, 327–345.
37. Kervaire, M. (1960). Some nonstable homotopy groups of Lie groups. *Illinois Journal of Mathematics, 4*, 161–169.
38. Kristensen, L., & Madsen, I. (1967). Note on Whitehead products of spheres. *Mathematica Scandinavica, 21*, 301–314.
39. Lang, G. E. (1972). Evaluation subgroups of factor spaces. *Pacific Journal of Mathematics, 42*(3), 701–709.
40. Lee, K. Y., Mimura, M., & Woo, M. H. (2004). Gottlieb groups of homogenous spaces. *Topology and its Applications, 145*, 147–155.
41. Liulevicius, A. (1962). *The factorization of cyclic reduced powers by secondary cohomology operations*. Memoirs of the American Mathematical Society (Vol. 42). Providence, American Mathematical Society.
42. Lundell, A. T. (1966). The embeddings $O(n) \subset U(n)$ and $U(n) \subset Sp(n)$, and a Samelson product. *Michigan Mathematical Journal, 13*, 133–145.
43. Mahowald, M. (1965). Some Whitehead products in $\mathbb{S}^n$. *Topology, 4*, 17–26.
44. Mahowald, M. (1967). *The metastable homotopy of* $\mathbb{S}^n$. Memoirs of the American Mathematical Society (Vol. 72). Providence, American Mathematical Society.
45. Mahowald, M. (1977). A new infinite family in ${}_2\pi_*^s$. *Topology, 16*, 249–256.
46. The Mathematical Society of Japan (1987). *Encyclopedic dictionary of mathematics*. Cambridge: MIT.
47. Matsunaga, H. (1961). The homotopy groups $\pi_{2n+i}(U(n))$ for $i = 3, 4$ and 5. *Memoirs of the Faculty of Science Kyushu University A, 15*, 72–81.

48. Matsunaga, H. (1963). Applications of functional cohomology operations to the calculus of $\pi_{2n+i}(U(n))$ for $i = 6$ and $7, n \geq 4$. *Memoirs of the Faculty of Science Kyushu University A, 17*, 29–62.
49. Mimura, M. (1964). On the generalized Hopf homomorphism and the higher composition, Part I; Part II. $\pi_{n+i}(\mathbb{S}^n)$ for $i = 21$ and 22. *Journal of Mathematics of Kyoto University, 4*, 171–190; (1965) *4*, 301–326.
50. Mimura, M. (1966). Quelques groupes d'homotopie métastables des espaces symétriques $Sp(n)$ et $U(2n)/Sp(n)$. *Comptes Rendus de l'Académie des Sciences Paris Series A-B, 262*, A20–A21.
51. Mimura, M. (1967). The homotopy groups of Lie groups of low rank. *Journal of Mathematics of Kyoto University, 6–2*, 131–176.
52. Mimura, M., Mori, M., & Oda, N. (1975). Determination of 2-components of the 23 and 24-stems in homotopy groups of spheres. *Memoirs of the Faculty of Science Kyushu University A, 29–1*, 1–42.
53. Mimura, M., & Toda, H. (1963). The $(n + 20)$-th homotopy groups of n-spheres. *Journal of Mathematics of Kyoto University, 3–1*, 37–58.
54. Mimura, M., & Toda, H. (1964). Homotopy groups of $SU(3)$, $SU(4)$, $Sp(2)$. *Journal of Mathematics of Kyoto University, 3–2*, 217–250.
55. Mimura, M., & Toda, H. (1964). Homotopy groups of symplectic groups. *Journal of Mathematics of Kyoto University, 3–2*, 251–273.
56. Morisugi, K. (1987). Metastable homotopy groups of $Sp(n)$. *Journal of Mathematics of Kyoto University, 27–2*, 367–380.
57. Morisugi, K. (1991). On the homotopy group $\pi_{8n+4}(Sp(n))$ and the Hopf invariant. *Journal of Mathematics of Kyoto University, 31–1*, 121–132.
58. Morisugi, K., & Mukai, J. (2000). Lifting to mod 2 Moore spaces. *Journal of the Mathematical Society of Japan, 52–3*, 515–533.
59. Mukai, J. (1966). Stable homotopy of some elementary complexes. *Memoirs of the Faculty of Science Kyushu University, A20*, 266–282.
60. Mukai, J. (1982). The S^1-transfer map and homotopy groups of suspended complex projective spaces. *Mathematical Journal of Okayama University, 24*, 179–200.
61. Mukai, J. (1988). A remark on Toda's result about the suspension order of the stunted real projective space. *Memoirs of the Faculty of Science Kyushu University, A-42-2*, 87–94.
62. Mukai, J. (1991). On the attaching map in the Stiefel manifold of 2-frames. *Mathematical Journal of Okayama University, 33*, 177–188.
63. Mukai, J. (1995). Note on existence of the unstable Adams map. *Kyushu Journal of Mathematics, 49*, 271–279.
64. Mukai, J. (1997). Some homotopy groups of the double suspension of the real projective space RP^6. *Matematica Contemporanea, 13*, 235–249.
65. Mukai, J. (2001). Generators of some homotopy groups of the mod 2 Moore space of dimension 3 or 5. *Kyushu Journal of Mathematics, 55*, 63–73.
66. Mukai, J. (2008). Determination of the order of the P-image by Toda brackets. *Geometry and Topology Monographs, 13*, 355–383.
67. Mukai, J., & Skopenkov, A. (2004). A direct summmand in a homotopy group of the mod 2 Moore space. *Kyushu Journal of Mathematics, 58*, 203–209.
68. Nomura, Y. (1974). Note on some Whitehead products. *Proceedings of the Japan Academy, 50*, 48–52.
69. Nomura, Y. (1980). On the desuspension of Whitehead products. *Journal of the London Mathematical Society (2), 22*, 374–384.
70. Nomura, Y. (1983). Self homotopy equivalences of Stiefel manifolds $M_{n,2}$ and $V_{n,2}$. *Osaka Journal of Mathematics, 20*, 79–93.
71. Nomura, Y. (2002). *A letter to J. Mukai on* $[\iota_n, \nu_n^2]$, 6 June 2002
72. Oda, N. (1976). Hopf invariants in metastable homotopy groups of spheres. *Memoirs of the Faculty of Science Kyushu University A, 30*, 221–246.

73. Oda, N. (1979). Unstable homotopy groups of sheres. *Bulletin of the Institute for Advanced Research of Fukuoka University, 44*, 49–152.
74. Oda, N. (1986). Homotopy theory and related topics. *Advanced Studies in Pure Mathematics, 9*, 231–236.
75. Ôguchi, K. (1964). Generators of 2-primary components of homotopy groups and symplectic groups. *Journal of the Faculty of Science University of Tokyo, 11*, 65–111.
76. Paechter, G. F. (1956). The groups $\pi_r(V_{n,m})$ (I). *Quarterly Journal of Mathematics Oxford Series (2), 7*, 249–168.
77. Pak, J., & Woo, M. H. (1997). A remark on G-sequences. *Mathematica Japonica, 46–3*, 427–432.
78. Rutter, J. W. (1968). Two theorems on Whitehead products. *Journal of the London Mathematical Society, 43*, 509–512.
79. Rutter, J. W. (1969). Correction to "Two theorems on Whitehead products". *Journal of the London Mathematical Society (2), 1*, 20.
80. Serre, J. -P. (1953). Groupes d'homotopie et classes groupes abéliens. *Annals of Mathematics, 58*, 258–294.
81. Siegel, J. (1969).G-spaces, H-spaces and W-spaces. *Pacific Journal of Mathematics, 31–1*, 209–214.
82. Steenrod, N. (1951). *The topology of fibre bundles*. Princeton Mathematical Series (Vol. 14). Princeton: Princeton University Press.
83. Sugawara, M. (1953). Some remarks on homotopy groups of rotation groups. *Mathematical Journal of Okayama University, 3*(2), 129–133.
84. Thomeier, S. (1966). Einige Ergebnisse über Homotopiegruppen von Sphären. *Mathematische Annalen, 164*, 225–250.
85. Toda, H. (1962). *Composition methods in homotopy groups of spheres*. Annals of Mathematical Studies (Vol. 49). Princeton: Princeton University Press.
86. Toda, H. (1963). Order of the identity class of a suspenson space. *Annals of Mathematics, 78*, 300–325.
87. Toda, H. (2003). *Unstable 3-primary homotopy groups of spheres*. Study of Econoinformatics (Vol. 29). Himeji: Himeji Dokkyo University.
88. Varadarajan, K. (1969). Generalised Gottlieb groups. *Journal of the Indian Mathematical Society, 33*, 141–164.
89. Whitehead, G. W. (1946). On products in homotopy groups. *Annals of Mathematics, 47*, 460–475.
90. Whitehead, G. W. (1950). A generalization of the Hopf invariant. *Annals of Mathematics, 51*, 192–237.
91. Whitehead, J. H. C. (1953). On certain theorems of G.W. Whitehead. *Annals of Mathematics, 58*, 418–428.
92. Whitehead, G. W. (1978). *Elements of homotopy theory*. Berlin: Springer.
93. Wu, J. (2003). Homotopy theory of the suspensions of the projective plane. *Memoirs of the American Mathematical Society, Chapter 5, 162*(769).

Index

© Springer International Publishing Switzerland 2014
M. Golasiński, J. Mukai, *Gottlieb and Whitehead Center Groups of Spheres, Projective and Moore Spaces*, DOI 10.1007/978-3-319-11517-7

Printed in the United States
By Bookmasters